Erik Dhont Landscape Architects

Works 1999–2020

Erik Dhont
Landscape Architects

Works 1999–2020

HATJE
CANTZ

From Garden Art
to Landscape Architecture
and Back Again

Michael Jakob

Garden and landscape are not synonymous—far from it. While the garden has been around "forever," landscape is regarded by most specialists as a typically modern phenomenon. The situation is complicated by the addition of a third element, landscape architecture. This young discipline began to emerge in the late eighteenth century and was, from the start, both related to and in competition with garden art. But there was also another complication in the field and in the theoretical works so popular at the time: Landscape architects regarded themselves as rather more than garden designers. Humphry Repton (1752–1818), an early pioneer of the new discipline, identified himself on his business card as

Humphry Repton's business card engraved by
Thomas Medland, c. 1808.

a "landscape gardener"; his chief occupation may have been designing gardens in the new picturesque taste, but he resisted being seen as a mere garden planner. Meanwhile, others who were part of garden design's long tradition began to turn their attention to the new aesthetics of landscape. In the second half of the eighteenth century, that age of revolutions, the combination of garden art and landscape spread and established itself. For the first time in history, the landscape garden found itself in the limelight if only for a brief, intense period. As the correspondences between the arts gained recognition, the garden emerged as a place where the various aesthetic forms came together—poetry, painting, music, architecture, and nature interacted to create the apotheosis of art. During these years, classical aesthetics (not yet known as such, because Alexander Gottlieb Baumgarten's *Aesthetica*, which gave the discipline its name, would not be published until 1750) fell into decline. For centuries, gardens had not been considered beautiful unless they corresponded to a formal ideal based on established values such as harmony, proportion, completeness, or clarity. The sublime and then the picturesque, both of which would permeate Enlightenment history, negated and transcended this traditional notion of beauty, while at the same time acting as ultimate, more powerful versions of it. To this complex scenario we should add the triumph of landscape. Not the pictorial representation of landscape—which in the fifteenth century provided both the term and the phenomenon we now refer to as "landscape"—but the experience of the world as landscape: a mental representation of a slice of nature.

The complexity outlined in this brief overview is not limited to the historical context. The events of the eighteenth century laid the foundations for phenomena, conflicts, and possibilities that are with us to this day. Can someone be both landscape architect *and* garden designer in the twenty-first century? Are contemporary garden design and landscape projects underpinned by a recognizable professional figure, a tradition, and a solid foundation? Or are we—that is, those who design and create both garden and landscape projects—playing by ear?

Landscape Architecture

Let us begin with a brief and entirely heuristic summary of the three fundamental entities under consideration: landscape architecture, garden art, and the experience of landscape. Landscape architecture was first identified and defended as a discipline in about 1789. Repton in England, and some years later, Jean-Marie Morel in France, were trailblazers in a new profession that presented itself as more ambitious, more interdisciplinary, and more open than garden art.[1] Landscape architects work on a grand scale, aiming for land transformation that goes beyond circumscribed space. They work like engineers—and sometimes with their help—and must coordinate the efforts of several different occupations. They leave behind them the closed areas of the sites that are to be developed and set to work in a rhizomatic fashion on the spaces in between; in this way, the new discipline claims and occupies streets, squares, parks, housing estates, urban fringes—all those spaces not given over to architecture. Once landscaping has been established as a recognized practice, however, problems begin to arise, not least the questions raised by the term "landscape architecture." In what way is the work *architectural*? Is the landscaper not, in fact, the architect's adversary?

The history of the profession offers ample proof. From Repton to Frederick Law Olmsted and all the great twentieth-century and contemporary figures, the relationship between landscape and architecture is a complex one, resembling a battleground where competing worldviews vie with each other. On a diachronic level, the relationship is one of a polar opposition between two ways of interpreting space: at one end of the spectrum, the landscaper slavishly submits to the good will of the architect, while at the other, we see genuine dialogue and, more recently, the triumph of landscape architecture over architecture itself. What is at issue here are not different theoretical possibilities or simple disagreements between two distinct areas of expertise. On-site interventions in the name of architecture (this, at least, is one implication of the term "landscape architecture") have different connotations from those carried out under the banner of landscape. Close examination, then, reveals the landscaper as the architect's *other*—an otherness that architecture itself, a private client, or public body can profit from by pursuing richer and more complex project developments.

The situation becomes even more confused when landscape architecture enters into competition with urban planning. Given that a significant number of landscape projects are carried out in or near an urban space and that cities now dominate the world, there will inevitably be a clash between two philosophies: on the one hand, the systemic, analytical, and objectivist tradition of the urban planner and, on the other, the aesthetic, ecological, and subjectivist approach of the landscaper.

A final complication is the astonishing fact that the discipline of landscape architecture has never written its own history. Landscapers have been so busy transforming the world and asserting themselves against architects, engineers, urban planners—and recently sometimes ecologists, too—that they have neglected both the theory and the historiography of their discipline. We are thus dealing with a practice that undeniably has significant merits and many works to show for itself—but that hardly ever shows them, or only occasionally, as an individual achievement. No one, not even those who are really interested in the subject, truly seems to know landscape architecture. We can of course list a few names: Repton, Olmsted, Carl Theodor Sørensen, Gerrit Eckbo, Dieter Kienast —but we have trouble placing them on the spatiotemporal map of an ensemble, of a coherent collective practice. The main consequence of this is a glaring lack of visibility. Whereas buildings tend to assert themselves with powerful architectural gestures and never cease to talk about themselves, landscape projects often remain silent, crushed into submission by the massive and supposedly more "useful" structures that dominate the view.

It is therefore clearly necessary to increase the visibility of landscape architecture in order to lay the foundations for future projects. Showing, discussing, and drawing attention to Dhont's work is one way to do this, but more is at stake than a presentation of the creative universe of the Brussels-based garden-designer-cum-landscape-architect, whatever that may mean; these preliminary reflections are an attempt to find an answer. To map out Dhont's itinerary is also to contribute, albeit modestly, to the as yet unwritten history of landscape architecture.

Garden Art

This effort to trace Dhont's trajectory could equally well take the more classical and solid phenomenon of garden art as its starting point, since the bulk of his work seems in keeping with the more accessible dimension of gardens. Here, again, though, brief reflection is enough to reveal a situation of bewildering complexity. Even more than landscape architecture, garden art is characterized by a precarious identity. Gardens, in other words, are in a constant state of flux. A garden is never identical to itself, so to speak; it is always in motion, often to the point of becoming unrecognizable. The idea that we can capture the essence of a garden by analyzing its layout or grammar is an illusion, because what is crucial to the experience of a garden is the "lived" moment of discovering the garden itself. Gardens are complicated narratives with more or less stable protagonists that evolve within a framework of constant change. The second aspect undermining the idea of the garden as an easily identifiable entity is a stylistic one. When we think about gardens, where are we coming from, and where are we heading? The history of gardens in Europe is, of course, well known,

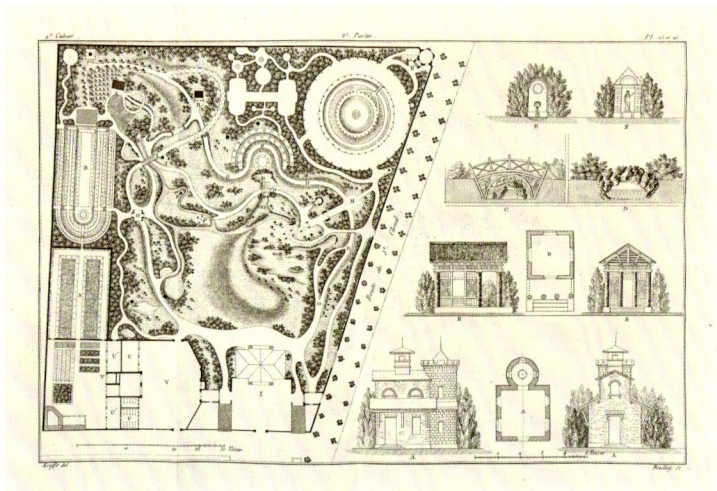

Jean-Charles Krafft, *Plans of the Most Beautiful Picturesque Gardens in France, England and Germany, and of the Edifices, Monuments, Fabrics, etc., which Contribute to Their Embellishment, of Every Kind of Architecture, such as Chinese, Egyptian, English, Arabian, Moorish, etc.* (Paris, 1809).

at least when it comes to major styles and currents, which for lack of a better term we identify as "Italian," "French," and "English" (or "Anglo-Chinese," "picturesque," "irregular," and "scenic"). Yet the history of gardens in the West has been characterized by eclecticism ever since the first half of the nineteenth century. Whether in Gabriel Thouin's magnificent plates (*Plans raisonnés de toutes les espèces de jardins* [Reasoned Plans of All Kinds of Gardens], Paris 1820) or in the pages of Gustave Flaubert's genius posthumous work *Bouvard et Pécuchet*, what dominates garden art is mixture, pastiche, and the endless replication of itemized forms and solutions. With no style taking precedence over any other, it seems that anything goes in garden art. This culminates in a grotesque hodgepodge of the kind described in Flaubert's ironic vision:

> It was an alarming sight in the gloaming. The rock, mountainlike, filled the lawn, the tomb was a cube in the middle of the spinach, the Venetian bridge a circumflex accent over the beans, and the hut beyond a large black blot, for they had set fire to the thatched roof to make it more poetic. The yews, in the shape of stags or armchairs, stretched away as far as the lightning-struck tree, which extended crosswise from the arbor to the bower where tomatoes hung like stalactites. Here and there a sunflower spread its yellow disc. The Chinese pagoda, painted red, looked like a lighthouse on the mound. The peacock's beaks, caught by the sun, reflected the glow, and beyond the lattice fence, now free of its boards, the dead flat country edged the horizon.[2]

Throughout the last centuries, garden art has oscillated between the formal and the informal; sometimes closer to architecture, sometimes to nature. Thus William Robinson's "wild garden" was overthrown by the return of the geometric garden, which in turn was challenged by the "green cathedrals" and "vagabond plants" of the late twentieth century. This to-ing and fro-ing of the garden's aesthetic grammar confers an extraordinary degree of latitude and freedom on designers, but it is also a source of fundamental uncertainty. If almost everything is possible and almost everything has already been done and meticulously documented, then either relativism sets in—and everyone does as they please, because, as Joseph Beuys said, *jeder Mensch ist ein Künstler* (everyone is an artist)—or we are condemned to the eternal return of forms belonging to the past. There is, then, a crisis not only in landscape architecture, but also in garden art, and this is an issue that should be addressed by all designers, present and future. Forging a path of one's own is difficult for garden designers and landscape architects alike.

Landscape as Experience

The third phenomenon of our initial triad, lived landscape, has no simple solutions to the dilemma either. We are truly living in an age of total landscape, in a society where landscape seems to have a secure place, even—as evidenced by the European Landscape Convention—on a legislative level.[3] Landscape today is monitored, protected, represented, and admired. But when we say *landscape*, what exactly do we mean? Landscape comes, as we know, from the tradition of European painting. For centuries, the terms *landschap*, *paysage*, *paesaggio*, and *paisaje* meant no more or less than a painted landscape, and the French word *paysagiste* meant not landscaper, but landscapist—a painter of landscapes. The history of this genre—the pictorial representation of a slice of nature—began in the fifteenth century and went on until the nineteenth, when the slow but inexorable "death of landscape" began.[4] With such artists as Paul Cézanne,

Alexander Cozens, *A New Method of Assisting the Invention in Drawing Original Compositions of Landscape*, Plate 4 (London, 1785).

Claude Monet, Piet Mondrian, and Jackson Pollock, the logic of landscape found itself in crisis, because viewers, confronted by representations that eluded conventional spatial orientation, were no longer able to identify the images before them as landscapes. In modern art, the conditions of possibility for a painted landscape no longer seem to exist. Viewers can no longer get their bearings *within* the fictional space of the work they are looking at. The alternative is, at best, only a kind of *post-landscape*—a deconstructed landscape à la Gerhard Richter, for example.

In the eighteenth century, however, the new mental representation—i.e. landscape experienced by someone on site—begins to prevail over its artistic representation. We have learned to recognize and identify a great variety of landscapes (sublime, picturesque, Romantic, melancholy, bucolic, and so on) and we do so almost automatically. But the identity of these landscapes—that is to say, the lived landscape that I see and create through my gaze—is nevertheless problematic. It is certainly I who sees this landscape, here and now. But as soon as I have seen it—and perhaps even as I am looking at it—*this* landscape is liable to disappear at any time. All perception of landscape is ephemeral, and there are really only *moments* of landscape. What happens to my liberty at such precious moments when, as I become aware of the bit of nature before me, a landscape comes together only to fall apart again? The identity of a lived landscape, of the landscape we experience, depends on cultural factors. We learn to identify and focus on certain kinds of landscape; we develop a liking for nocturnal or Romantic, rural or Alpine landscapes. In this way, the acts that create landscapes do not entirely belong to me, and my gaze tends to be controlled rather than free.

Hieronymus Rodler, *Eyn schön nützlich büchlin ...*
(A Fine Useful Booklet ...) (Simmern, 1531).

These seemingly overly theoretical considerations become particularly important when landscape is connected to the two other areas of this inquiry. When specialists—garden designers or landscape architects, say—work on a project that will inevitably have considerable aesthetic impact, are they free in what they do, or are they influenced by the diktat of a certain way of looking? Take the picturesque. This aesthetic category, invented by British theoreticians and popularized by such figures as the Reverend William Gilpin or Uvedale Price, has become a defining language or style for a significant proportion of the gardens, parks, and landscapes of the last almost two and a half centuries worldwide. The appreciation of a picturesque scene does not come to us naturally; it must be learned. So what effect does the continued importance of the picturesque have on the creative freedom of those who design gardens and draft landscape projects? I ask these questions in the conviction that the hybrid nature of landscape architecture, the eclecticism of twenty-first-century garden art, and the very ambiguity of the phenomenon of landscape provide fertile and challenging ground for contemporary designers.

The World in Writing

In light of these thoughts, the one thing I am not going to do is force Dhont into a pigeonhole—and, given the range of his work, it would hardly be possible anyway. It suits me very well that he should move back and forth between the virtuosity of garden art and the complexity of landscape architecture, between formal elegance and botanical knowledge—that he should sometimes navigate shifting ground. It is even proof that he has a lot to say.

How does he make such a good job of taking a clear stand when he is beset by uncertainty and under pressure from clients? How does he work in a field with so few points of reference, a discipline with no established foundation or dogma? The answer lies in his undeniable cultural sophistication. Dhont and others like him who leave their mark on the world rely heavily on culture and, more precisely, on the instruction of books. In the all-digital age of the computer and the internet, this may seem like an anachronism. In fact, there is nothing more up-to-date or useful than a book; only hard copy can compensate for the lack of a global approach that is typical of landscape architecture and its relationship to garden art.

The relationships between garden art, landscape as a phenomenon, landscape architecture, and the world of printed text are extremely rich. Repton, the pioneer of landscape design, conceived, planned, and "sold" his projects with the help of his famous "Red Books."[5] These meticulously produced precious red-bound volumes had their place between the two poles of his work: the exploration of an existing site and the design of an inspirational future site. In many cases, the books have survived, although the completed projects disappeared long ago. In the Red Books, Repton collated brief texts, maps, sketches, and "before-and-after" watercolors that compared the existing state of a site with its future state. His clients traveled through the book as if through the site itself, passing from detail to detail, imagining the site in its redeveloped state. Drawing, writing, and arranging the elements in a page-by-page narrative was not, however, the sole prerogative of the landscape gardener; the proprietors fortunate enough to have their estates redeveloped by Repton were themselves urged to put the results into words and pictures, some of which also appeared in one or another of the Red Books.

Humphry Repton, *Design for the Pavilion at Brighton* (London, 1808).

This strong link between garden art and books runs very deep. Garden history is also the history of texts in gardens and, more generally, of their textual quality. A garden—certainly an expressive garden—is a complex enunciation, with its own grammar, figures of speech, and rules. Crossing a garden or walking in it is tantamount to an act of speech. Landscape, on the other hand, seems at first glance asemantic and formal: we do not read it; we look at it. Upon reflection, though, it becomes clear that here, too, the primacy of language is not without effect, because constituted landscapes tend to be *recognized* landscapes. Our attention is caught by landscapes that resemble those we have already seen. It is too often forgotten that the historical development of the landscape—a centuries-long process—was underpinned by a plethora of publications: the descriptive poems of Albrecht von Haller, James Thomson, and Jacques Delille, the novels of Jean-Jacques Rousseau and Ann Radcliffe, the pedagogical works of the aforementioned Gilpin and Price, the literary wanderings of William Wordsworth and Samuel Taylor Coleridge and the travel accounts and sundry guides that taught first art lovers and then an ever-growing public to identify landscape patterns: elements arranged in a formal, picturesque, or Romantic manner.

Linguistic primacy asserts itself in a similar fashion in the field of garden knowledge, starting with the identification of plants. We see only what is identifiable; we see *through words*. That is why we cannot create "interesting situations," to use the term proposed by the Marquis de Girardin, without an awareness and profound knowledge of the textual—a knowledge acquired chiefly through reading books. The important gardens of the past, and particularly the great Italian gardens, in which Dhont has always shown a keen interest, are essential models in this respect. Bomarzo, for example, is a key to understanding his work, because this sixteenth-century garden north of Rome, sometimes designated very superficially and simplistically as a "park of monsters," is an enunciation as complex as a poem. Bomarzo *speaks* and it does so precisely by moving away from the traditional grammar of Renaissance gardens. The sequence of sculptures and follies in the park, their effects of surprise, their "graphic" appearance in the mysterious grove, the inscriptions with their magnificent patina—in this place designed by Vicino Orsini and Pirro Ligorio everything is arranged in such a way as to make the dialogue between nature and art as expressive as possible. In addition to this, Bomarzo is an eminently literary ensemble, where statues recalling the epic poem *Orlando Furioso* are interspersed with trees and shrubs that tell tales of their own: the laurels, for example, relate the inconclusive love story of Daphne and Apollo; the poplars, the tragic fate of Cygnus.

The Aesthetics of Atmosphere

The most important lesson, however, that Dhont has learnt from Bomarzo—and similarly important Renaissance gardens, like the Boboli gardens in Florence—is that of atmosphere. This mention of a category as apparently vague as atmosphere may seem surprising. But in almost every encounter with Dhont's projects, it is precisely atmosphere that strikes you first and demands explanations—and in this context the great Italian gardens provide a perennial paradigm.

For too long, the focus in garden art has been on the aesthetics of production. We analyze the designs, we inventory the plants, we observe the interaction between buildings and nature, we focus on the relationship with the clients, assuming that the "secret" of a garden can be easily translated into clear and distinct concepts. This approach is like treating aesthetics as a theory concerned principally with objects. And yet, if there is any work of art whose identity goes beyond the factual and objective, it is the garden in all its aesthetic valency. What attracts our attention in a garden is that *je ne sais quoi* inherent in the term

The connection between architecture and nature, open space, and groves.

atmosphere. Of course, the atmosphere of a garden—its mood or *Stimmung*—depends on the elements that go to make it up. The topography, the arrangement of the plants, the paths, the follies, the views—all these come together to produce an *effect* on us that is not immediately explicable. Atmosphere is tangible and direct, but because it is the result of so many different factors, it is also enigmatic and elusive.

From its origins in the mid-eighteenth century until the twentieth century, aesthetics neglected atmosphere, focusing instead on the material aspect of a work of art. It was not until the insights of Georg Simmel and the recent theories of Gernot Böhme that philosophy began to concern itself with atmosphere as the determining factor in our encounter with art. Now, when we assess an architectural project, which could be a building or an open space—a garden, square, or street—the most important consideration is ultimately that of atmospheric quality. "Diffuse and ethereal, atmosphere (according to Böhme) is less a matter of any objective property than an ambiance that suffuses things, *coloring* the way we experience them."[6]

What gives one garden an atmosphere that is immediately beguiling, while most others leave us indifferent? Atmosphere may elude definition, but that need not prevent close analysis. The atmosphere of Bomarzo or of the landscapes designed by Dhont is the result of a specific poetics, in the sense of the Greek term *poiesis*, or craftsmanship. Such works are imbued with atmosphere because of the sensitivity with which they present nature. At Bomarzo, the interplay of light and shadow, with its many nuances, produces surprising effects and makes us look *differently* at, say, a large tree, a statue, or a series of vases. In Dhont's gardens, the arrangement of hedges creates structures that blend plenitude and emptiness, vegetable and mineral, vertical and horizontal, giving the overall conception a rhythm of its own immediately apparent to the visitor, a *Gestalt* reminiscent of musical form (something that further eludes description). It is thus possible—and Dhont's work is tangible evidence of this—to produce not atmospheres, but the condition of their possibility. For this, the garden designer and landscape architect must add two further strings to his or her bow: knowledge of scenography and the graphic arts. (It is worth remembering that Dhont started his career as a graphic artist.) The other indispensable factor in creating a meaningful, immediately palpable atmosphere is the practice of sketching. Dhont's projects are born of his sketches—of the intelligence of the hand that arranges the various elements differently from the analytical brain or the large number of computer programs available to specialists.

A Poetics of the Sketch

There is a drawing by Repton in which he portrays himself with a piece of paper in one hand and a pencil in the other. With the pencil he points to a sketch on the paper: the origin of his art. All that I have done, all that I can do, comes from there, he seems to say. For garden designers and landscapers, drawing is a kind of writing; it is their way of putting on paper a design that will later come into being on the ground. Relying on a sketch or a series of sketches also has temporal implications: it takes a considerable amount of time to fine-tune an idea in a

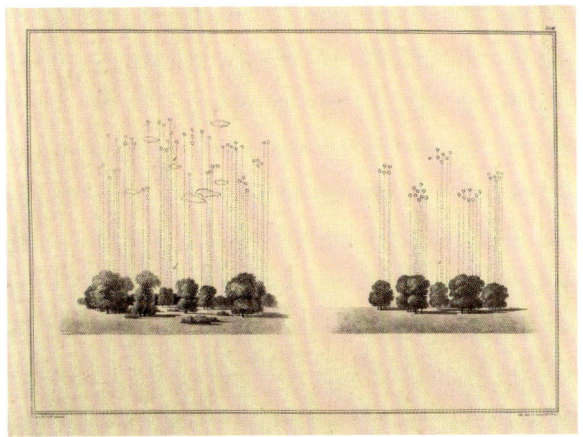

Herrmann von Pückler, *Andeutungen über Landschaftsgärtnerei verbunden mit der Beschreibung ihrer praktischen Anwendung in Muskau* (Hints on Landscape Gardening with a Description of their Practical Application in Muskau) (Stuttgart, 1834).

sketch that is still far from the final draft. This time encourages reflection and, most importantly, it allows for incorporating the accidental, the surprising, the unexpected. A designer who puts him or herself at the mercy of a sketch commits to an open-ended, polysemous journey; a sketch conveys more than its author is aware at the moment of putting pencil to paper. The typology of sketch, which I have in mind does, however, have a specific origin. At the crucial moment when a project begins to take shape, the experiences gathered by its author are indispensable; in this context, it makes sense that regular use is made of *topoi*, both concrete and literary. A sketch that comes into being gradually as a solution to a specific problem is thus a palimpsest, the tangible product of a culture of landscape and gardens. This culture helps to create an evolving grammar and a repertoire of forms that cross the boundaries between one era and another, between garden designers and landscape designers. A recurring element in Dhont's work, for example, is the formal pattern—the more or less regular arrangement of rectangular or square blocks, which become garden compartments, hedge structures, and so forth. This formal or geometric style is not exclusive and can be combined with informal solutions from all areas of garden art. It is part of a tradition that extends from the Roman *ars toparia* to the Italian and French gardens of the Renaissance and has seen recent revival in the—very diverse—works of Dieter Kienast, Dan Kiley, Roberto Burle Marx, and Carl Theodor Sørensen. Dhont's formal language, however, has its origin not only in cultural memory (meaning that each major project implicitly or explicitly alludes to a whole series of previous gardens), but also in the poetics of the sketch, which I have been describing. The geometric objects and the manner of their arrangement, with all that this implies in the way of play of light and distance and scale, come into being on paper and retain the freshness and immediacy of a sketch, even once they have been translated onto the ground.

Garden Design and Landscape Projects Today

Although Dhont's work includes landscape architecture projects in the public domain, the bulk of it is related to garden art. Given this, it may seem strange to talk of a "politically committed" approach. Is it possible to design elegant and sometimes almost inaccessible gardens while at the same time pursuing a course toward actively transforming the status quo? I should like to reply in the affirmative: The focus of Dhont's work lies precisely in the creation of "another world." This "world" often consists of little things: the patches of color provided by a series of bushes, the remarkable "graphism" of the bark of a tree, the play of light through foliage. If we look for the common denominator of the projects conceived and implemented by Dhont, we soon discover that his work obeys the general logic of beautifying the world.

Aesthetics, as we know, came into being in a period marked by a crisis of beauty. The traditional rationale that associated pleasure with classical values such as clarity, harmony, and completion proved inadequate in an age where that sense of allure, the *je ne sais quoi*, came not only from works of art, but also from objects that were irregularly formed, strange or downright brutish. Beauty suddenly found itself in competition with other categories, to the point where the *nicht mehr schönen Künste* sometimes won out over the traditional models. At the same time, nature, which had previously been regarded as ugly and non-aesthetic, was, for the first time in European history, promoted to the rank of primary source of aesthetic emotions. Despite its new appeal and recognition, however, nature was not "beautiful"; it was sublime or picturesque—it was, in a sense, more than beautiful.

The state of "supermodernity"[7] in which we live allows the beautiful to co-exist with competing aesthetic categories such as the sublime in the strong sense of the term, the picturesque, the Romantic, the ugly, and so on. In traditional artistic disciplines such as painting, architecture, sculpture, and even cinema, beauty now seems out of date. To speak of the survival of the beautiful in gardens or landscape projects implies that such places belong to an aesthetics of the past.

But there is nothing backward-looking about contributing to the world's beautification. In the current situation, which is marked by the increasing banalization and standardization of the built environment, any attempt to make the environment in which we live more habitable is an act of resistance, as is any attention bestowed on a plot of land, however small—a garden or a square. The distinctive combination of plant presence (founded on an excellent knowledge of botany) and formal elegance (often the result of the way those plants are arranged) creates a new, oxymoronic reality in Dhont's work. This new form of beauty, so necessary today, is oxymoronic because it submits to an order, or rather a Gestalt, that combines the vegetable *and* the mineral, plenitude *and* emptiness. It is a beauty very different from the classical ideal, because it embraces hybrid forms and surprising combinations. It never ceases to refer metonymically to nature in general, while at the same time accepting the artificiality that characterizes it as a work of art.

The magic of Dhont's projects lies in their ability to restore meaning to a place and to give root, momentarily or permanently, to those who enter it. In an age when banalization and standardization too often prevail, and when the care of our lived environment is by no means guaranteed, Dhont's approach seems therapeutic, if not essential.

1. See Stephen Daniels, *Humphry Repton: Landscape Gardening and the Geography of Georgian England* (New Haven, 1999); Joseph Disponzio, *The Garden Theory and Landscape Practice of Jean-Marie Morel* (New York, 2000).

2. Gustave Flaubert, *Bouvard et Pécuchet* (Paris, 1909), p. 61. Trans. Imogen Taylor.

3. See Michael Jakob, *L'émergence du paysage*, (Gollion, 2004); *Temps et paysage* (Gollion, 2007); *Le paysage*, (Gollion, 2008).

4. See Kenneth Clark, *Landscape into Art* (London, 1949); François Dagognet, ed., *Mort du paysage?* (Seyssel, 1982).

5. André Rogger, *Landscapes of Taste: The Art of Humphry Repton's Red Books* (London, 2007).

6. Emmanuel Alloa and Céline Flécheux, preface to Gernot Böhme, *Aisthétique—pour une esthétique de l'expérience sensible* (Paris, 2020), p. 9.

7. See Marc Augé, *Non-Places: An Introduction to Supermodernity* (London, 2009).

— Spaces & Scenery

Wherever we are, we apprehend space through our bodies. With all its senses, the human body perceives the characteristics of the spaces in which it lives and moves—spaces made of volumes, textures, colors, and light, but also sounds and smells, and even temperature and air quality. As a landscape architect, Erik Dhont creates spaces and composes views. His parks and gardens are living spaces to occupy, traverse, and view—either from somewhere else or from within the space itself.

On a practical level, these spaces are always defined areas. They are plots of land—large or small, public or private—which the landscape architect arranges, forms, and sculpts in accordance with the owners' wishes and the site's existing features. In any given space, Dhont might completely reshape things or simply add a discreet personal touch. The result will feel so natural that, in studying the photographs in this book, many readers might ask what has been (re)done. The answer, in most cases, is everything. Or almost everything.

Whether it involves reshaping a site's topography to smooth uneven terrain (579 Tervuren, above), using topiary to create open-air rooms around a house (227 Geneva) or planting lush undergrowth as a backdrop for a swimming pool (511 Mons), Dhont's work consists of creating scenery on several levels. In (re)drawing these spaces, he plays on the interaction between topographical particulars and architectural and vegetal elements, their structure and texture, inclusion or suppression.

Then come the details: positioning the right plant in the right place, with the right outline, in the right quantity.

Some of the tools Dhont uses are centuries old and traditional to the art of landscape: the use of earthworks, land leveling, compartments, hedges, and topiary; basins and fountains, steps and walls; crafting raw plant material such as trees, bushes, grasses, flowers, mosses, and ferns ... a rich world of plants.

All these elements—the landscape architect's props and scenery, as it were—play on our senses to create a space, a stage set, distinctive to the places conceived and (re)structured by Dhont. How does he influence our perception of the places he creates? What are the threads of his narrative?

The spaces in question all have one or more functions, of course, depending on the people who inhabit or pass through them. But they are also affected by their position in the landscape—whether they are in the country or city center, and how they are situated in relation to, say, a private house, a housing estate, a business, a public space, a square, a road, a river, or the edge of a wood. On the other hand, when we speak of spaces and scenery, we evoke a construct that is not only highly visual, but also metaphorical. The magic is twofold: The landscape is both stage and spectacle. In his projects, Dhont makes us look at places in a new way. The clearly defined limits of a site rarely correspond with what we see. We perceive things on

3

1/
579 Tervuren (BE)
Undulating open space accompanies the view
to the house.

2/
643 Brussels (BE)
A lap pool in a landscaped environment,
seen from the house.

3/
080 Lier (BE)
Undergrowth—a space to contemplate the microcosm;
a scene that could be an ode to nature.

many different levels: a distant view of the ocean, for example (279 São Miguel), or of hills (554 Saint-Rémy-de-Provence), a view along a waterway (680 Coburg), a small enclosed space (617 Brussels), or a microcosm planted along the edge of a path (080 Lier, above).*

Dhont steers our gaze in a particular direction—toward a building or a landscape—or guides our steps, suggesting a walk, inviting us to wander through topiary (467 Bruges) or find out what is at the end of a pergola (301 Ohain). The art of a landscape architect is to design a space that surprises, enchants, and comforts—by creating a variety of tableaux, a series of little worlds, but also by orchestrating more functional places: a car park (280 Roeselare), a pathway between levels (228 Delvaux), or simply a bench to sit on and enjoy the scenery (274 Lanaye).

Dhont believes that one of the main challenges to the landscape architect is to create a tension between spaces, to position them appropriately in relation to one another in such a way that all the elements are interconnected—as if someone had cast a net over the entire landscape and pulled it tight. That said, those elements, striking though they may be, are limited to the strictly necessary. Perspectives, frames, and itineraries are suggested rather than dictated, and enough space is left for the sky, the surroundings, the unpredictable side of nature.

In Dhont's designs, the solid bodies are balanced by empty spaces: implied shapes and outlines, like negatives, which play their own part in the way we perceive the landscape, making us live to the rhythm of *his* landscapes, along paths, between topiary plants, past flowerbeds, under a tree, on a flight of steps, or at the edge of a pond. We move within these spaces, walking, going up and down, stopping, turning back and immersing ourselves. In this way we take possession of the places—while at the same time succumbing to their game.

* Attentive readers will notice that the examples cited in each of the seven chapters are not confined to the landscape projects pictured there. The themes of the individual chapters are those that all of Erik Dhont's gardens draw upon and echo—each in its own way, depending on the specifics of the site, but all bearing the imprint of their creator. Every project is presented from a particular angle in order to underline certain aspects, but each of those aspects can also be found in the other gardens.

Two gardens have been joined to create one urban garden. Elegantly sculpted yews arranged in a kind of colonnade form the focal point. A mature walnut tree and an old fountain have been preserved. The garden is scattered with architectural features that represent the encounter between the mineral and the vegetal worlds. The harmony between built and living elements can be seen in the combination of botany and architecture on the narrow path between two walls—and in the swimming pool, which is designed to be invisible from the house and whose material substance is one with the environment.

The existing vegetation is balanced with the clipped shapes of the topiary. A composition of chiaroscuro is created in the undergrowth with climbers and ground cover.

Existing vegetation: *Juglans regia*
Evergreen vegetation: *Buxus sempervirens, Taxus baccata*
Shade-loving vegetation: *Asplenium scolopendrium, Dryopteris affinis*

This project surrounds an imposing eighteenth-century manor house with a view of Mont Blanc. The first phase was to restore the main courtyard. A square bench was installed in the center to recreate the original layout and act as a meeting place. The second phase was to introduce a contemporary note to the historical part of the park, giving new life both to the original French garden and the more recent English landscape garden. A walkway of monumental yews leads from the center of the building toward an old cedar and overlooks a seventeenth-century canal. The sloping ground and the topiary's slanting edges create a tension between the different parts of the park.

Existing plantings, such as the horse chestnuts, underwent regeneration. Wisteria and a climbing rose frame the facade.

Trees: *Acer palmatum, Aesculus hippocastanum, Juglans regia, Prunus yedoensis, Sophora japonica, Tilia cordata*
Clipped shrubs: *Carpinus betulus, Taxus baccata*
Groves: *Cornus controversa, Corylus avellana, Hamamelis mollis "Pallida," Osmanthus sp.*
Clusters of flowers: *Anemone hyb. "Honorine Jobert," Delphinum sp., Lupinus sp., Paeonia suffruticosa, Papaver orientale sp., Scabiosa atropurpurea, Thalictrum delavayi*
Bulbs: *Allium giganteum, Allium sativum, Eremurus himalaicus, Narcissus thalia*

228 Overijse (BE)

The new garden creates a link between a cottage-style house from the early twentieth century and a lower-lying park. The supporting walls have been camouflaged with rhododendrons. A walkway of clipped yews forms a harmonious link between the old building and the new one, visible from the upper terrace, which also offers a view of the topiary. Close to the terrace the yews are cut low, just tall enough to compensate for the difference in height. But as the ground falls away, the topiary becomes taller, the tallest trees reaching a height of four meters, so that anyone walking through the garden experiences a variety of sensations. Meanwhile, the spaces between the yews are an incitement to play and relax.
The garden is composed of three levels: the clipped shrubs (*Taxus baccata*), the lawns, and the trees.

Trees: *Acer platanoides, Carpinus betulus, Cedrus atlantica, Fraxinus excelsior, Juglans regia, Quercus robur*
Shrubs: *Hamamelis × intermedia "Diane," Hydrangea paniculata, Hydrangea quercifolia, Viburnum opulus, Viburnum sieboldii, Taxus baccata*

511 Mons (BE)

The challenge of this project was to save the existing wooded park around the new building, while making it botanically more interesting. A limestone path leads from the house to the swimming pool across a lush herbaceous bed stretched out beneath pines. The walkway continues under the existing oaks and beeches diversified by a new bed of spring- and autumn-flowering shrubs such as Japanese camellias, rhododendrons, and a field maple, which contrast with

a background of evergreen foliage like holly. In the undergrowth, a wide range of ferns and shade-loving plants, such as wild garlic and wood anemones, provide constant variation.

"Cushion" of greenery: *Buxus sp., Chaenomeles nivalis, Ilex crenata*
Undergrowth: *Luzula sylvatica "Onderbos," Polypodium vulgare, Polystichum acrostichoides*

617 Brussels (BE)

This garden in the center of Brussels was the result of a
conscious decision to create an exuberant urban jungle
where nature has free expression—contrasting, in its
diversity, with the house's austere and minimalist interior.
The garden has been planted with a wealth of flowering
shrubs such as *Pieris* and *Enkianthus*. Over time, the walls
will be covered in scented climbers, and a marvelous
collection of ferns will carpet the ground. Along the
edge of the terrace, exotic plants with finely cut foliage,
such as spikenards and Japanese maples, complete the
ensemble. The upper part of the garden, which adjoins
a park, is densely planted with evergreens.

Trees and shrubs: *Enkianthus campanulatus, Rhodendron
praestans, Fatsia japonica, Azal-eodendron sp. "Mauve"*
Perennials and ferns: *Asplenium scolopendrium,
Polypodium vulgare, Dryopteris wallichiana, Epimedium
youngianum "Niveum," Helleborus argutifolius*

— Elegance & Wilderness

When we speak of wilderness, we tend to think of untouched nature, of flora and fauna left to their own devices. In today's world, such wild spaces are increasingly rare, drastically transformed by human activity. Elegance, on the other hand, is considered the work of humanity: totally artificial and inherently cultural, long sought-after and constantly improved upon; the result of a quest for aesthetic pleasure combining expertise, history, and ideally, impeccable taste.

That is, to a certain degree, true of most historic gardens that date from a time when landscape was only partially cultivated and still included large swathes of land undisturbed by human intervention. In those days, tamed nature was perhaps the only form of nature in which humans could feel safe.

Today the situation is largely the reverse, and the human species has become a danger to a significant part of the natural environment. We like to "go green," turning to nature to collect ourselves and find some kind of peace far from the noise of urban and industrial areas. What we tend to forget in these oases of greenness is that, in a landscape inhabited by humans, everything is in fact artificial and cultural, influenced more or less directly by human activity. Similarly, in a park or garden, everything is the work of the landscape architect, even those parts that may appear "natural" or "wild."

A landscape created by humans seems to us easily recognizable as such when it uses regular forms, geometric patterns, architectural elements, and cultivated or exotic plants, especially ones that have been visibly trimmed or sculpted. If, however, a landscape appears unordered and perhaps even overgrown and chaotic, we perceive it as natural. But what about landscapes that have been groomed into perfect irregularity?

Erik Dhont often completely redesigns a landscape (274 Lanaye) or makes use of an existing structure such as mature trees (511 Mons). We may have the impression that the landscapes have always been there, or that the flora has taken hold naturally, but this is not the case. Some spaces seem wild, but are not. Such wildernesses are the work of Dhont's hand, and the natural effect is just that—an effect. Dhont stakes out the spaces and fills them with carefully chosen plants.

Indeed, Dhont puts the contrast between wilderness and regularity to conscious use. By planting lush perennial foliage at the foot of large undulating screens of trimmed yews (080 Lier) or setting clumps of buttercups in a meadow of tall grasses traversed by a mown path (280 Roeselare), he plays on our uncertainty. Perhaps his intervention is, in fact, minimal? Just enough to make the place habitable and agreeable? No. His intervention is total. It is, however, in dialogue with the place—and with what we call nature.

As precise as Dhont's designs are, they always leave a certain amount of scope for nature. He creates circumstances that allow for the newly introduced flora to harmonize with the existing ecosystem. The same goes for the fauna; the new landscape will be a biotope for them, too. Dragonflies, crayfish, hedgehogs, titmice, and owls are "anticipated" and welcome (680 Coburg).

3

1/
450 Brussels (BE)
In its most natural form, wilderness can have a certain elegance, even in the heart of an urban garden.

2/
202 Wetteren (BE)
Micro-planting: a source of botanical richness that enhances the architecture.

3/
258 Malibu (US)
Natural planting around a sunken tennis court highlights and integrates the architectural ensembles while providing players with a backdrop of wilderness.

The beauty and elegance that we see in a landscape at this stage are also artificial. They are the product of our human gaze, our deeply cultural vision of nature. The wilderness is not beautiful; we merely consider it beautiful and perceive its elegance and sometimes even its sublime force.

The devices that allow Dhont to play with the contrast between highly designed spaces and apparent wilderness have a clear aesthetic goal, of course, but they also serve an ecological purpose. Design and wilderness feed off the other and, in Dhont's work, neither can exist without the other. For him, artistic creation must be consistent with our natural environment. He strives for equilibrium between the creative intervention of humans and the self-organization of nature, although the exact configuration is up to him.

One way of maintaining or reestablishing this equilibrium concerns the ecological choices regarding ecosystems in a given landscape. It consists in the conscious creation of biotopes and even microclimates that favor the development of certain flora and fauna. Particular attention is given to plants that are local or capable of developing in the new environment—in the climatic conditions, the type of soil, the degree of exposure to wind and sunlight, and so on. By leaving enough space for land to develop naturally, Dhont reintegrates humans into the habitat of which they are an intrinsic part. We do not live outside the environment, but we can shape and humanize it—cultivating it while at the same time sparing a thought for nature's microcosms and macrocosms. The less a plot of land is tended, the more plants and wild animals will inhabit it.

Wilderness provides elegance with quality and depth. The fact that greenery and nature have become valuable assets in our post-industrial world makes the wild features of a landscape fascinating and beautiful. In Dhont's work, the dialogue between the artificially shaped and the natural creates tension and rhythm. The spaces he designs allow us to observe the wild side of nature (536 Aalst), to walk around in it (280 Roeselare), or even to see it enhanced by the stark contrast with architectural or sculptural structures (351 Damme).

By making the best use of a site's existing conditions and turning the constraints of a place into advantages, Dhont lends his aesthetics to a landscape—an aesthetics dictated by natural conditions, but adapted to the way of life of a human being who chooses and modifies their biotope.

Clear contours and accessible paths are naturally more comfortable and suitable for humans. In his landscapes, Dhont thus creates visual contrasts that create an aesthetic tension while aiming for ecological harmony.

280 Roeselare (BE)

This garden was created on the site of an old farm, and bricks from the demolished buildings were used to construct large, ruin-like structures that give the place an atmosphere of intimacy. Between the bricks, a variety of ferns and perennials were planted, which will over time heighten the effect of wildness. The surrounding land is wet, criss-crossed by canals and interconnecting ponds; in the spring it is rich with buttercups and lilies. This reinforces the natural look of the place, while making it possible to channel rainwater all over the site.

A series of basins regulates the flow of water. The vegetation makes it possible to observe the landscape without being seen.

Wetland vegetation: *Alnus glutinosa, Quercus robur, Ranunculus repens, Phragmites communis*

173 Brussels (BE)

This former wasteland in the municipality of Schaerbeek was transformed into a 5000 square meter park. The plans for Reine Verte Park show concrete supporting walls arranged in steps for a sculptural effect. A waterfall and bushes will soften their overpowering appearance and attract small birds and butterflies. Despite the considerable difference in heights, all parts of the park remain easy to access thanks to the design of two paths. The first takes a quick and direct route using ramps and steps; the second meanders and offers more intense interaction with the site. Planting includes trees, chosen for their translucency, and colorful lower-level vegetation such as hazel and hydrangea shrubs. A "jungle walk" was also designed for small children.

Trees: *Tilia cordata, Quercus robur, Prunus avium, Sophora japonica, Magnolia soulangeana "Brozzoni"*

080 Lier (BE)

This project consisted in restoring and maintaining an early nineteenth-century park surrounding a neoclassical building from the same era. Subtle changes were made to restore the original character of the place while recreating the gardens' structure and botanical richness. Large clumps of rhododendrons and azaleas were transplanted to create timeless views and spaces. New conifers were planted, along with a collection of magnolias and other flowering trees and shrubs. A small bridge and the garden's first fabriques were restored. A spectacular walkway of cutting flowers was laid out between undulating hedges of *Taxus baccata*, making for some dynamic scenery.

The undergrowth near the topiary was chosen and arranged in collaboration with the gardener. The microrelief was introduced without disturbing the surroundings.

Trees and shrubs: *Calocedrus decurrens, Magnolia fraseri, Rhododendron ponticum, Rhododendron luteum, Taxodium distichum*
Topiary: *Taxus baccata*, in combination with undergrowth vegetation
Microrelief: *Acer osakazuki* and various mosses

351 Damme (BE)

This project involved adding several new features to the estate of a renovated farm while preserving the privacy of the area around the house. The existing limes and poplars that blend in with the surrounding countryside were thinned and clipped; the canals and lake surrounding the house were cleared and their banks restored. Large clumps of shrubs and new trees were planted, contributing to the rural integration of the place and its natural enhancement. Around the house, hornbeams, oaks, and hollies are combined with a variety of *Rosaceae*, and beds of perennials are edged with box. The client's

art collection is displayed in a landscape that seems almost to have been left in a wild state.
The ditches were enlarged to create habitats for avifauna. The bridges and walkways are made of the same wood species as those found in the landscape.

Trees and shrubs: *Carpinus betulus, Corylus avellana, Cornus mas, Ilex aquifolium, Juglans regia, Quercus robur, Syringa vulgaris "Charles Joly"*
Topiary: *Carpinus betulus*

584 Oostkamp (BE)

This landscape project involved restoring a château park that had been expanded several times over the years. First the differences in level were evened out. Then an area was made for the swimming pool in the garden's most lavishly planted area, with vegetation designed to be particularly attractive in the summer. The combination of rhododendrons and woodland plants creates a perfect frame for a view of the park and looks different from every angle. Each part of the park has its own colors, mainly thanks to seasonally flowering shrubs with striking autumn foliage. The existing pond was renovated. Its banks were cleared and redeveloped as ecologically as possible, to allow for natural and aesthetic evolution over the next sixty to eighty years.

Trees: *Nyssa sylvatica, Magnolia denudata, Taxodium distichum, Pinus sylvestris*
Undergrowth: *Convallaria majalis, Dryopteris wallichiana, Dryopteris cycadina, Epimedium pubigerum, Galium odoratum, Helleborus niger, Anemone sylvestris, Polypodium vulgare*
Groves: *Corylopsis spicata, Osmanthus heterophyllus, Viburnum × bodnantense, Viburnum × burkwoodii*
Azaleas: *Azalea "Charles Rogier," Azalea "La Surprise," Rhododendron "Durante Alle"*

274 Lanaye (BE)

The aim of this canal project on the Dutch-Belgian border was to design a new system of locks and weirs and construct a landscape by reusing the gravel and silt removed during the canal's construction. A series of new earthworks was made on the existing flat ground, and the changes in elevation created inaccessible areas, providing natural habitats, including wetlands, for wild fauna. A walkway makes it possible to observe, from a distance, these new islands inaccessible to humans. Resting places encourage strollers to enjoy the views over the countryside, where the flat ground meets the rocky cliffs along the canal.

Trees: *Alnus glutinosa, Betula pendula, Fraxinus excelsior*
Wetland perennials: *Iris pseudocorus, Sagittaria sagittifolia, Sparganium erectum*

— Texture & Structure

A landscape is a place that affects all our senses. In this book, the visual dimension dominates; we cannot smell the scent of the lime blossom or the cut grass, or hear the sound of the grasses in the wind or the animals that inhabit the gardens. Nor can we bite into the fruit in the orchards, or touch the slabs of polished granite, the bark of the trees, or the moss growing at their roots. But we can see the texture of all these things and imagine what they feel like.

One point to remember when analyzing or interpreting a work of landscape architecture is the importance of plants to the landscape architect. They are one of a garden's raw materials, along with architectural elements such as wood, stone, brick, cement, terracotta, glass, and metal. All these materials offer an enormous range of possibilities, but when it comes to plants, the variety is almost infinite. The botanical world provides a plethora of plants of different shapes, colors, sizes, and textures, to name only the visual aspects. These include mosses, ferns, succulents, grasses, and all kinds of flowers and trees.

Landscape architects use plants for their aesthetic and visual properties, as they might use sculptural or ornamental features. The same is true of the other components they employ—paving and edging stones, a basin's mosaic tiles, fountains, canals, and garden furniture. The difference is that these things are inert, while plants are living organisms that constantly change and develop. Plants also provide the perfect complement to inert

objects: perennials and bulbs between a checkerboard labyrinth and geometric yew topiary (389 Deauville), a cobbled path crossing an herbaceous area (536 Aalst), or a serpentine box hedge, light-leaved trees, stone, and ponds (431 Brussels).

The result is a pronounced sculptural character, but also strong tactility. The plants' visual aspect makes us aware of their feel and texture. At the same time, plants have an effect on their environment. The silhouette and texture of a tree, a piece of topiary, or a cluster of flowers influence the movement of the wind and other air currents around them, as well as the degree of moisture retention. They also influence the way in which the light is broken or filtered by the foliage or on the sides of the topiary. This same light will be reflected differently depending on whether it falls on water, a white wall, a wall of volcanic stone, or on bromeliads or agaves (279 São Miguel).

All of this is scenery on a small scale. The texture depends on the characteristics of the materials used (whether vegetable or mineral), but also on how they are used. Clumps of grasses, meadows of tall grass, or a closely mown lawn—the smaller the element, the more it lends itself to contrast. A larger or taller element can be varied by altering its natural form, cutting or shaping it. Clipped trees are very important to the work of Erik Dhont, who regards them as creative elements in outdoor architecture. There is a centuries-old tradition of topiary in European garden art and many examples can be found in Italy, France, and Belgium.

3

1/
961 Oostduinkerke (BE)
The forms and textures of the vegetation allow it to create intimate spaces in the dunes' landscape structure.

2/
720 Graty (BE)
The vegetation emphasizes the seasons, thanks to the weather's effect on the plants' colors and textures.

3/
795 Vandoeuvres (CH)
Combinations of plant texture are an inexhaustible source of inspiration in constructing a space.

Topiary is scenery on a large scale; it creates green "rooms" and lends individuality to a garden. Dhont has some distinctive examples in yew and box, peopling his gardens like silent and motionless inhabitants (080 Lier, 227 Geneva, 228 Overijse, 301 Ohain, 351 Damme, 389 Deauville, 431 Brussels, 467 Bruges, 554 Saint-Rémy-de-Provence, 691 Antwerp).

The topiary's angles and curves produce various and sometimes surprising patches of light that change throughout the day, according to the weather and seasons. Although highly artificial and abstract, the trees' shapes allow for the close observation of atmospheric phenomena that are often not easily perceptible.

The sense of proportion and interest in structure that Dhont brings to his landscapes were developed and refined during his graphic art studies, even before he became a landscape architect. His studies of typography and books gave him an eye for design and scale. Because no two environments are alike, each requires a new approach as well as different plants and materials; Dhont also gives his own particular touch to each.

The diversity of textures—both those present in nature and those produced by working with materials—permits the landscape architect to create a particular structure for every landscape. Dhont creates what he calls tension in a place by arranging various elements of a landscape until an equilibrium has been established. If the space is overloaded, the tension disappears. It is crucial to know not only how to bring energy and density to a site, but also when to stop. An overload of elements is counterproductive and disturbs the balance of a landscape. The right harmony is found when the forms and materials are arranged in a way that seems self-evident. As if nothing could be more natural.

389 Deauville (FR)

This eighteenth-century Norman estate, divided into several sections, has been radically transformed. A checkerboard labyrinth forms a link between the two main buildings, while preserving a certain distance. The regular rhythm of large paving stones gives way to tall yew topiary, bulbs, and perennials. The slender forms of the topiary open up the space, breaking free from the rules of classical structure. The dark yews contrast with the softer green of rippling perennials. To be immersed in this new abstract landscape is to enjoy an aesthetic pleasure that is constantly renewed as different plants bloom. The boundaries have been elevated to enhance this effect, and a series of mounds arranged along the inner edges.

Accompanying flora: *Actaea simplex, Aquilegia vulgaris, Gaura lindheimeri "Whirling Butterflies," Helianthus decapetalus, Leucanthemum superbum "Becky"*
Topiary checkerboard: *Taxus baccata*

431 Brussels (BE)

The front garden, originally a space that opened onto the road, was closed off for greater privacy. This move was reinforced by planting tall trees, clipped into organic shapes, on either side of the access path, and creating a sequestered walkway. White flowering dogwood connects the front garden with the back garden, which originally comprised a lawn and a pond with a rock garden. A snaking box hedge encloses the rock formation, its sculptured appearance thrown into relief by the meticulous lawn that is visible from the house. At the bottom of the garden, this hedge becomes the backdrop to a wealth of seasonally changing species.

Trees: *Sophora japonica, Prunus yedoensis*
Groves: *Cornus "Eddie's White Wonder," Hamamelis intermedia "Westerstede," Philadelphus × virginalis "Virginal," Syringa vulgaris sp.*

196 Itterbeek (BE)

This garden, next to a modern house, was designed for a family with five children. Walking around it means engaging in a game of shifting lines and surfaces, structured by clipped topiary and planted banks. The soil that was dug out to make the swimming pool was used to create two rows of three banks along a sunken walkway. The outsides of the banks are sown with grass; the insides are planted with perennials and spring bulbs. The banks give the flowers height, making them visible from the house. A terrace in blue shale stone runs the length of the house and overlooks the garden. The lawn cuts into the edge of the terrace, making it feel less overpowering.

Orchard: *Malus domestica "Belle fleur double," Malus domestica "Boskoop," Prunus domestica "Mirabelle de Metz"*
Flowering bank: *Fragaria vesca, Helenium autumnale, Myosotis vulgaris, Scabiosa columbaria*

536 Aalst (BE)

The landscape project for Aalst Crematorium was conceived as the contemplation of an inaccessible landscape, a space to provide support to those separated from a loved one. It comprises a static garden made up of spaces for contemplation, a large body of water with hygrophilous vegetation, conducive to meditative walks, and a long, highly molded valley that extends the view from the inside of the building to the outside. These shifts in terrain make for a sense of calm and peace when entering the crematorium site—an effect that

will increase over time, thanks to the limited accessibility and natural development of the designed spaces. The contours and plantings help to shield the place from the sight and sounds of the road.

On-site hydroseeding: *Campanula rapunculus sp., Cardamine pratensis, Echium vulgare, Helleborus viridis, Knautia arvensis, Omphalodes verna, Reseda lutea, Sanguisorba officinalis*

— Climates & Seasons

Every landscape is part of a specific site under the influence of distinct geological and climatic zones. A landscape is, so to speak, inherently "local." Seasons and climates are not easily influenced; they are natural criteria to which humans must adapt. Landscape architects are of course affected by these changeable and sometimes unpredictable factors. They design their works of landscape accordingly. This is a challenge, but also an opportunity. The living material in a garden imparts an element of fascination, intensified by the variability of external factors.

Because of their cyclical nature, the seasons are associated with a certain regularity. This aspect —at once repetitive and dynamic—makes it necessary to observe certain rules, but also allows for the creation of an infinite variety of scenes that change by the minute, season, decade, or half-century. Colors, shapes, and textures evolve and succeed one another. Erik Dhont likes to create a perfect ensemble, choosing plants that will thrive and taking into account the changes brought about by the seasons, the weather, the time of day, and the presence of flowers and birds. A place gets its character not only from a particular design and layout, but also from a succession of unique moments and the perfect blend of the planned and the unpredictable.

Natural changes take place on several levels —night and day, sun and rain, fog and frost, evening summer breeze and winter storm, dry cold and humid heat, budding and blooming. These changes affect the visual and olfactory atmosphere; they affect our behavior, and that of the plants and animals. Dhont takes all these criteria into account and remains conscious of them throughout his orchestration of a landscape work.

A garden that is exposed to frequent sun and powerful light enjoys expressive effects of light and shade, and intense smells. It also settles into a moment of peace at the end of the day when the sound of cicadas gives way to the gurgle of water in the gargoyle, connected to a tank to give the impression of abundant water, like in an ancient Oriental garden (554 Saint-Rémy-de-Provence).

Climate therefore affects not only a garden and its development, but also how we make use of the space. Depending on the climate, we have different needs. We structure our days differently—and as a result, we structure our landscapes differently, too. This is the origin of a number of age-old traditions.

Water is by definition vital in any landscape and often an important factor in the choice of plants. It is also a valuable resource that gains importance even in places that are little affected by drought. In areas of frequent rainfall, that are close to the water table, or are prone to flooding, water management can pose a significant challenge—or it can be used to engineer a specially designed landscape with pools and aquatic flora (600 Louvain-la-Neuve).

In many of his works, Dhont renders climate visible, making it a feature by incorporating it into his design. By accentuating existing conditions, he aestheticizes them, sublimates them. The natural circumstances become a driving force of landscape.

3

A landscape exposed to the wind and surf has to contend with immense forces, for example. The resulting strength and robustness are enhanced by the nature of local volcanic stone. Dotted with clouds carried by air masses, the sky above changes constantly. Below, tropical plants, protected from storms by low walls or their own ground-growing nature, gleam in every shade of green, and bright blooms bring splashes of warm color (279 São Miguel).

With global warming, certain factors are becoming less predictable, increasingly tending to extremes that pose new challenges. It is therefore all the more necessary to understand why and how certain soils and climates make it possible to use some plants rather than others: gardens that can survive frost and several weeks of drought—or even gardens that do not require human intervention (487 Antwerp).

The desire to understand and make practical use of seasonal and climatic cycles can be found in what is probably the grandmother of all gardens —the vegetable patch. Here, in a patchwork of shapes and colors, space meets time to provide aesthetic and culinary pleasures all year round. Architecture can help the garden withstand the winter; the presence of water in the form of a fountain can give pleasure to the ears—it is certainly indispensable for growing things to eat (729 Brussels).

Dhont sees landscape as a dynamic entity and visualizes this dynamism in time as well as space to give it an aesthetic and poetic dimension. His gardens are designed to be beautiful throughout the year; he carefully plans and varies the emphasis as the seasons change. It is a long-term performance that makes use of cycles. For example, by choosing species of the same genus that flower at different times (varieties of camellias, in this case), Dhont makes one walkway perform month after month, an ongoing firework display with constantly varying shapes and hues. It eventually cedes the stage to lush, glossy leaves for the rest of the year (301 Ohain).

Here again, space meets time. Or rather, time modulates space. For a place never looks the same twice; it is changed by the sunsets and moons, the banks of mist and scatterings of rain that cross and caress it. And any plant can be positioned so that it is the star of the landscape at a particular moment, depending on season, climate, and fate— not to mention the skill of the landscape architect, which always presents something of a challenge to serendipity.

This garden is situated on the top of a cliff surrounded by the ocean. Once a vineyard, it was structured with low walls and terraces to protect the vines from the wind. These structures were redesigned with slopes and indentations, creating visual abstraction and giving the garden a more contemporary look. A swimming pool in local lava stone is protected on one side by two high walls. Avenues and terraces make use of the same stone, differently cut. The landscape comprises a series of themed gardens: a pleasure garden with a large collection of proteas, a jungle garden evocative of lush tropical vegetation, an ornamental kitchen garden, and an Atlantic garden with bromeliads, agaves, and agapanthus.

Trees and shrubs: *Beaucarnea recurvata, Plumeria rubra, Brugmansia versicolor, Callistemon viminalis, Metrosideros excelsa, Tetrapanax papyrifera, Protea neriifolia, Schefflera actinophylla, Dracaena draco*
Succulents: *Aeonium glandulosum, Aloë plicatilis*
Ferns: *Asplenium aethiopicum, Blechnum spicant*
Perennials: *Alpinia purpurata, Chasmanthe floribunda, Echium plantagineum, Thymus caespititium*

554 Saint-Rémy-de-Provence (FR)

The new garden of this old estate provides a bedrock for the château while echoing the historical landscape. The plantings of box, holly, and free-growing shrubs create a new kind of topiary garden and allow for a gentle transition to the surrounding countryside. Around the original beds of box, a garden of fluffy oaks and a botanical garden with a ground cover of spring bulbs and Mediterranean herbs have been planted. The new access road to the château prolongs the experience of being in the garden; it cuts across mixed plantings of olive trees and lavender and ends by skirting a rock formation at the edge of the estate. Water is used both for its visual and aural qualities.

Existing trees: *Arbutus unedo*
Topiary: *Feijoa sellowiana, Arbutus unedo, Phillyrea angustifolia, Pistacia lentiscus*
Individual landscape plants: *Arbutus unedo, Buxus sempervirens "Rotundifolia," Pistacia lentiscus*
Mediterranean perennials: *Achillea crithmifolia, Frankenia laevis, Gaura lindheimeri, Salvia scarlea, Salvia leucophylla "Figueroa," Phlomis italica, Rosmarinus officinalis "Sappho," Teucrium flavum, Thymus hirsutus*
Potted plants: *Plumbago zeylanica*

729 Brussels (BE)

The rhythm of the seasons begins in the earth where the first shoots form; later, we see (and taste) it on our plates. Thanks to the different flowering seasons and changing colors, the vegetation is in action on several levels, all year long. A solitary shrub has pride of place among the tubs of vegetables and gives shade to those in the center. Prunus trees add height to the garden, and hedges of *Carpinus betulus* structure the space by adding a sense of privacy. A fountain, forged by a blacksmith, provides for refreshment and relaxation, in addition to fulfilling more practical needs. The lapping water adds a musical touch to the overall effect, so that being in this garden becomes an experience to awaken the senses.

Herbs: *Coriander sativa, Hyssopus officinalis, Borago officinalis, Geranium macrorrhizum, Satureja montana*
Trees providing a backdrop to the vegetable garden: *Cydonia oblonga*

600 Louvain-la-Neuve (BE)

Water and climate. Nature and industry. The planning project at the AGC LLN headquarters is built around a natural approach to site development, which envisions the landscape itself as a sustainable solution to the needs and problems of surface water management. The sloping ground was extensively reshaped to blend in with the surrounding countryside's rolling hills. Alongside the parking lot beneath the building, an open space was preserved and carefully shaped in a series of water-retaining hollows. Over time, and with extensive management, wetland ecosystems will develop here, attracting orange-tip butterflies and dragonflies.

Wet grassland: *Cardamine pratensis, Deschampsia cespitosa, Luzula sylvatica, Panicum virgatum, Ranunculus ficaria*

— Time & Identity

How do we recognize a place? What makes a place unique? Identity, however pronounced, is paradoxically always defined in relation to something else. It expresses itself through characteristics of its own, in opposition to other identities. It is a means of demarcation. The identity of a landscape, moreover, is a snapshot, localized in space and time, tied to a particular moment and place. Identity seems so well-defined as to be immutable, whereas time by its very nature is fluid, associated with perpetual change. What does that mean for the identity of a landscape which, as we have seen, is by definition a work of life, time, flux, and change? Does time define a landscape by its past, present, and potential future? And what is the role of the landscape architect in all of this?

Erik Dhont is not interested in having a signature, or imitating historical styles. All that is tied up with too many constraints and restrictions. He does not believe that a landscape takes its identity from a particular style. Dhont does not attempt to make his mark on a site, but to understand its spirit. Especially when working on gardens that belong to cultural and historical heritage, his approach is to take that heritage into account; to honor it and allow it to develop in keeping with the spirit of the past. He creates a new identity with a contemporary vocabulary (680 Coburg). Reusing materials is another way of conferring a new monumentality to the gardens he designs (280 Roeselare).

Past and present exist alongside one another. There is no new without the old, and Dhont brings them together in a project that is made to be lived in the present, while awaiting the future development of plants and materials. Dhont is not only aware of this effect; he anticipates and encourages it in his works. Trees will grow (712 Oostduinkerke), topiary will thicken (351 Damme), flowerbeds will broaden (536 Aalst), pergolas will be overgrown with creepers (301 Ohain), moss will fill cracks between paving stones (437 Leuven). The present moment unites the past of a place with the landscape architect's vision.

One of Dhont's specialties is to revitalize historic sites with old stately homes, châteaux, or ornate farms. He combines the codes of historical heritage with a contemporary touch, while respecting the spirit of the garden which once existed (or may have done). He knows how to interpret that spirit and translate it into contemporary language using new forms, new materials, and surprising juxtapositions.

Dhont is famous for reinterpreting the age-old art of topiary, using shapes that are sophisticated but geometric and pared-down. He designs beds containing unexpected combinations of plants with unlikely colors and silhouettes, often lush and abundant, dabbed on as if with a paintbrush. But he also adapts the garden's functionality to our contemporary way of life and the needs and habits that accompany it.

Besides being tied up with its past and the possible presence of a historic building on the land, a landscape's identity depends greatly on the setting, whether it makes use of the surrounding view (279 São Miguel) or closes itself off from it (617 Brussels). Identity depends, too, on the topography

3

1/
731 Paris (FR)
Identity through plants and their materiality: different peony varieties adorn this garden.

2/
637 Paris (FR)
A garden's structure, materials, and composition can reinforce the spirit of a place. A pergola changes over time, existing to disappear beneath the plants it supports.

3/
485 Brussels (BE)
The choice of plants and use of space give a place an atmosphere of its own, in spaces conducive to the flourishing of life.

of a place—which may be created from scratch or exaggerated (258 Malibu)—and on the choice of forms and materials. These selections are made based on the sites in question and may result in specially designed paving stones (389 Deauville) or distinctive topiary (467 Bruges). Furniture and accessories, often designed by Dhont for a particular place, also play an important role. With their abstract but functional forms they are like sculptures, blending into the projects they are part of (227 Geneva). Historical identity is also present in the plants that have grown in a landscape for a long time—a natural heritage that Dhont honors and integrates into his projects.

Dhont never reuses an idea, because each is specific to a particular site. He does, however, like to come up with variations on ideas and explore new possibilities. To assure that he never repeats himself, he limits himself to three variations and designs each project individually.

Each design is conceived not only for a particular site but for the people who inhabit it. In this way, the users also contribute to a place's identity, by making certain choices to which Dhont responds with personalized solutions. This happens, for instance, when the redesign of a houses changes its relation to the outside space and the views (279 Brussels).

Landscape evolves over time; developing, maturing. Plants thicken it, as they take up more and more space; some materials age and gain patina —another way in which time contributes to a garden's identity.

Landscape architects make careful plans, projecting their work into the future and anticipating likely changes to a space once construction is complete. In their planting plans, they even include empty spaces that will be filled by time and architectural structures that will one day be covered in plants. Factoring in the temporality of space and raw materials means constantly redesigning and recreating space. The next challenge is to maintain the delicate balance of human intervention and burgeoning nature.

Sometimes, a place evolves more quickly than anticipated—when Dhont is asked to come up with a new version of a landscape he designed fifteen years previously. His job then is to reinvent himself—a self that, like his clients and their gardens, has changed and evolved.

279 Brussels (BE)

When this stately home was renovated, the owners opted for a contemporary design with large windows. This altered the spatial experience of the garden. The newly designed garden includes an impressive lake that continues the line of the sitting room. This lake was conceived as an interplay of positives and negatives; the scattered plantings and variously shaped and sized mineral features—some of them underwater—paint a complex design on the canvas of the water. The space is surrounded by a screen of evergreens, patterned with the graphic silhouettes of spindles and Japanese maples. When fully grown, these plants will be as tall as the garden wall, creating a kind of outdoor interior.

Trees: *Acer palmatum "Osakazuki," Carpinus betulus*
Aquatic plants: *Lysichiton camtschatcensis, Elodea canadensis, Potamogeton lucens, Stratiotes aloides, Cardamine pratensis, Epipactis palustris, Equisetum palustre, Nymphaea "Pygmaea Helvola"*

258 Malibu (US)

This site overhangs the ocean and is home to an art collection. Certain spaces were set aside for the artworks, which are arranged in a silent dialogue with nature.
To overcome significant differences in level, earth was removed in places, creating the illusion of a path cutting across the banks. Then new banks were constructed, using the excavated earth; this created spaces linked by walking trails. The site was planted with a natural effect in mind: a mix of indigenous plants on the mounds, many oaks, some additional plane trees to join the existing specimens, and a reduced number of eucalyptus trees.

A subtle lighting system in natural stone is reminiscent of moonlight.

Trees: *Aesculus californica, Cupressus macrocarpa, Quercus agrifolia, Lyonothamus floribundus*
Shrubs: *Arctostaphylos sp., Lupinus arboreus, Myrica californica, Quercus tomentella*
Perennials: *Achillea millefolium, Helianthus gracilentus, Perovskia atriplicifolia, Phlomis fruticosa, Salvia spathacea, Santolina chamaecyparissus, Thalictrum flavum*

The mineral dominates in this small urban garden, situated behind a period house that has been modernized by the addition of large windows. The idea was to create a ground covering of recycled materials: shale, ashlar, bricks. The pieces of shale were placed on their sides and the bricks split, giving them very sharp edges. A micro-planting of perennials and mosses between the stones creates a carpet of plants. Every morning, the garden is misty; a permanently damp atmosphere stimulates the spread of the mosses. Maples and standard wisterias add a vertical dimension to the space. Around the vegetal "carpet" are a terrace and path in blue stone. A border of climbers and box has been planted along the walls.

Trees: *Acer japonicum*
Climbers: *Hydrangea petiolaris*
Bulbs and Perennials: *Corydalis flexuosa, Helleborus foetidus, Narcissus "Thalia," Primula japonica "Alba"*

301 Ohain (BE)

This garden is made up of several walkways of flowers. A reinterpretation of the classical English border, the rose walk comprises hybrid perpetuals and musks, English roses, and botanic roses, planted in a precise rhythm. An elegant wooden pergola creates a play of shadows and acts as a picnic area. The walkway of 500 camellias arranged under a canopy of existing woodland trees makes visitors gasp with admiration—as does the topiary walkway, where osmantus, box, hornbeam, and beech contrast with flowering trees such as Japanese cherries. Visitors also stop beneath the palisade of gingkoes, to look at the big arcs of light filtering through the compact leaves.

Wisterias: *Wisteria sinensis "Consequa," Wisteria floribunda "Shiro Noda"*
Camellia walk: *Camellia japonica "Yours truly," Camellia japonica "Primavera," Camellia japonica "Joseph Pfingsten," Camellia japonica "Nobilissima"*
Focus in the camellia walk: *Magnolia denudata, Prunus × yedoensis*
Group of cherry trees in the vegetable garden: *Prunus "Accolade"*

709 Ghent (BE)

The original structure of this old manor garden—a raised courtyard and a sunken garden with a pavilion and a fountain—was preserved and some contemporary touches added. A blue stone rosette collects rainwater in the center of the paved courtyard, and roses and clematis covering the wall show the way to the lower garden. Here, a black brick path leads around the lawn. The restored fountain is edged with box topiary and small, distinctively shaped trees: cherries, magnolias, holly oaks. The borders are planted with hazels, roses,

and viburnum interspersed with perennials, creating an urban oasis and shielding the garden from view.

Solitary trees: *Magnolia loebneri "Merrill,"*
Prunus subhirtella "Automnalis," Prunus × yedoensis,
Quercus × turneri "Pseudoturneri"
Hedges: *Buxus sempervirens "Rotundifolia,"*
Fagus sylvatica, Ilex aquifolium, Ilex crenata,
Ilex altaclerensis "Camellifolia," Sarcococca humilis
Focus: *Hamamelis mollis, Virburnum × burkwoodii*

— Home & Escape

Home is a place of protection, where we can shelter —or simply take respite—from the outside world. Feared for its wildness, this world is excluded from the home because of its hostility or dense, urban nature. The garden is another place of protection, even etymologically: Most Indo-European words for garden come from a term connected with enclosure. Even paradise, that most ideal of all gardens, means a "compound or garden surrounded by walls"; the word is derived from the Persian *pairidaeza* (from *pairi*, meaning "around," and *diz*, meaning "wall, brick, or shape") and comes to us via the Greek *paradeisos*. The garden is thus an outside extension of the home, but always within a protected enclosure. Such distinctions recall those between the outside and inside worlds, and also, in a pinch, between public and private space, for home and garden are almost always private. Almost always. It is worth remembering that we can feel at ease in any place we happen to like—because it's calm, full of life, or beautiful. It doesn't matter whether it belongs to us.

It is when we are at home—or when we feel at home—that we can best escape. When we are protected, living in our own world, at our own rhythm, following our own rules. A garden allows us to retreat into another dimension within a microcosm that allows us to travel while staying put. When this microcosm is our own garden, we are in control; we decide what it should look like, just as we decide what kind of life should live there. We are free because we are at home.

In this private oasis, a landscape architect can design us a customized paradise, a microcosm of our own. In his landscape works, Erik Dhont puts a personal and contemporary spin on some archetypal gardens. The basic, utilitarian garden, cultivated to provide food and produce the vegetables, fruit, and maybe even cut flowers that we need, has never ceased to exist and is even making an urban comeback in the form of community gardens (799 Ruisbroek).

Escape within a garden can take many different forms. Being in a garden with a beautiful view makes it easier to transcend the everyday, and Dhont takes this to new heights by designing backdrops to breathtaking views of the Mediterranean (523 Cap-d'Ail), the North Sea and its dunes (712 Oostduinkerke) and Mont Blanc (795 Vandoeuvres).

In our own garden we are in control of a circumscribed space; a space that belongs to us and where we decide how things should look and who is admitted. Only certain inevitable rules of nature continue to hold. Even in a cultivated space, we have no control over the natural forces we have encountered in the previous chapters: climate, seasons, the passage of time. Part of the fascination of gardens lies in this lack of influence—in the very fact that a garden is made of living organisms that are subject to uncontrollable forces. Another feature of gardens is that they allow us to escape not only into another space, but into a temporal dimension to which we rarely have access—the present moment. In a garden, we live entirely in the present.

3

By coming face to face with a living being, a plant or animal, that lives only in the moment, we can find our center.

The large number of living beings that make up a garden make it seem like a living entity itself; there is something almost companionable about a garden.

Many biological phenomena, such as budding, flowering, and dormancy, happen of their own accord, but only at a certain moment, when the necessary conditions of temperature and light are met and the plants are "ready." It is not possible to be anywhere but the present in a garden, which has to be enjoyed *now*. Not sooner, however impatient you are; and not later, however busy you are.

Dhont aims to create gardens that astonish and delight. Often, they are living paintings. One of his credos is to create conditions that favor the unpredictable and its charms.

The millennia-old symbolic meaning of gardens plays an important part in our dreamlike retreats into their worlds. Archetypal and mythical gardens like the Garden of Eden, the Hanging Gardens of Babylon, or the Garden of Hesperides are a reminder that humans have been dreaming of beautiful gardens and imbuing them with positive associations since the dawn of time.

To this day, the living space of a garden thrills us and sets us dreaming, making us feel with all five senses, rooting us firmly in the present and reconnecting us with our bodies and with nature. It can be anything from an urban garden (467 Bruges) or a secret roof terrace (487 Antwerp) to the expansive grounds of a medieval château with beds of asters and anemones planted along a river (680 Coburg). Dhont's gardens are designed to provide us with a whole raft of experiences. They are exceptional places in which escape is played out over and over, without ever leaving everyday life.

680 Coburg (DE)

After decades of inertia, this park was revitalized with the aim of creating new harmony between the château and the surrounding countryside. The main prospect was completely redesigned and given new shape by diverting an existing stream. This contains a series of small islands planted with natural vegetation, which come into view as visitors walk along the water. Nature is center stage again. Various scattered additions such as a kitchen garden and a woodland walk bring diversity to the château's unity. Around the château, large limestone flags lend a contemporary touch. The plants growing in between the stones add a natural element to their imposing character.

Perennials: *Anemone hybrida* × *"Andrea Atkinson," Anemone hybrida "Carmen," Anemone hupehensis "Ouvertüre," Aster novi-belgii "Crimson Brocade," Aster novi-belgii "Marie Ballard"*
Interstitial plants: *Dianthus sp., Thymus sp., Achillea sp.*
Aquatic plants: *Lychnis flos-cuculi, Myosotis palustris, Equisetum japonicum, Filipendula ulmaria, Mentha aquatica*

795 Vandoeuvres (CH)

This landscaped garden of a 1920s chalet is situated outside Geneva, where the natural environment is all-important; swathes of forest grow right up to the entrance and cover the surrounding countryside, and there is an unrestricted view of Mont Blanc. This is a family garden, where gathering and spending time together are crucial. Some of the spaces are designed with this in mind; others are dedicated to solitude and contemplation in a setting where nature is ever-present. The different areas blend in with one another, combining orchards, woodland, and meadows. Rainwater is collected from the roof and channeled through the garden, allowing the lush and natural vegetation to thrive.

Flowering meadow: *Saponaria ocymoides, Silene vulgaris, Centranthus ruber, Echinops ritro, Centaurea montana "Grandiflora," Eryngium vulgare, Valeriana montana*
Undergrowth: *Allium ursinum, Anemone nemorosa, Dryopteris affinis, Geranium macrorrhizum, Onoclea sensibilis*
Groves: *Alnus glutinosa, Betula pubescens, Carpinus betulus, Frangula alnus, Pinus sylvestris, Sorbus torminalis*

799 Sint-Pieters-Leeuw (BE)

The ideal flower garden requires an understanding of botany and the effect of time. Here, the flexible nature of the fields creates a versatile space. The tomato greenhouse, surrounded by flowering fennel, forms a link between two adjoining vegetable gardens—one single-crop field and one field of herbs and cutting flowers. These fields are organized according to internal rotation schemes. At the center of the productive space, clumps of flowers and self-sown vegetation line the path. This garden is a place of experiment, knowledge, and wonder—a place to reflect on a future in which spontaneous natural development and agriculture are in equilibrium. The rotation scheme that structures the vegetable garden creates social dialogue through its changing appearance.

Annuals: *Calendula officinalis, Cosmos sp.*
Tubers: *Dahlia sp.*

712 Oostduinkerke (BE)

This dune landscape is devoted to walking and relaxation. The garden surrounds a sparely designed contemporary house that is partly enclosed to ensure privacy. Part of the garden was created by restoring the surrounding dunes, using chestnut trellises to support the sand. The dunes' contours were also reinforced by planting evergreens naturally adapted to the environment. The rhythm of the place creates different ways of interacting with the landscape—the paths, for example, bring out the raw character of the natural growth. Thanks to the evocative nature of the sea and the dunes, this project is imbued with a sense of escape.

Dune plants: *Ammophila arenaria, Acer campestre, Pinus sylvestris, Quercus robur, Quercus × turneri "Pseudoturneri," Rhamnus alaternus, Rosa arvensis, Rosa pimpinellifolia*

This garden, attached to an Art Deco house, comprises different ambiences conducive to escape, including an artificial grotto at the foot of the cliff. The materiality of this grotto contrasts with the vegetation growing there and the views of the sea. The harmony achieved by the seasonal diversity of foliage and colors makes every moment unique. A range of rock plants was reintroduced, notably *Plumeria rubra f. acutifolia, Tetrapanax papyrifer, Acca sellowiana, Agave victoriæ-reginæ, Philodendron monstruosum*, and various bromeliads. Botanical research and the use of space to bring out the richness of color, line, and texture have made it possible to conserve the place's character while emphasizing the Mediterranean context.

Trees: *Plumeria rubra f. acutifolia, Rhaphidophora decursiva, Tetrapanax papyrifera*
Ferns: *Asplenium nidus, Cyathea australis, Cyrtomium falcatum, Nephrolepis cordifolia, Platycerium bifurcatum*
Succulents: *Aeonium haworthii, Lampranthus coccineus, Sedum × rubrotinctum*

487 Antwerp (BE)

Situated in the center of the city of Antwerp, this roof garden was designed to make nature palpable in its purest possible form. An abstract rectangle cut into the roof catches the light and illuminates this living mineral garden, which takes the form of a library of stones on which mosses and lichens will grow. The library theme reflects the cultural world of the client. This outside library allows him to preserve a memory of his travels and bears witness to his expertise. Forming part of the house's private area, it is a space for meditation and a travel inventory, with room to add new stones.

— Research & Imagination

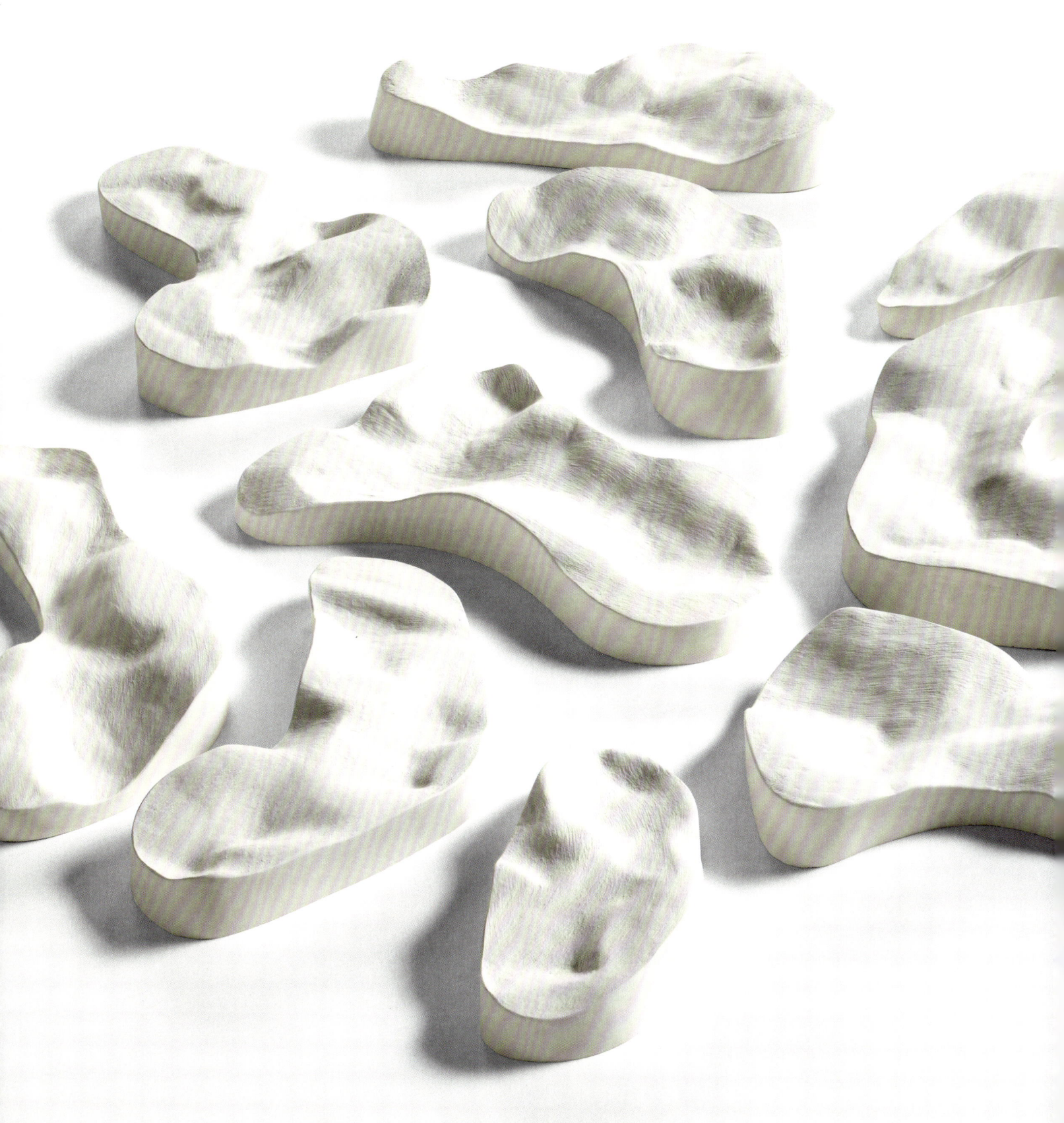

Erik Dhont's artistic research takes place on several levels. When he meets a client and is introduced to the site he is to work on, he begins, of course, by taking inspiration from the place and its history, the flora and the local culture. This sets in motion a creative process and a search for solutions to a specific problem.

But beyond his work on particular spaces and situations, Dhont is constantly looking for more general solutions. His approach is to reflect on subjects that interest him. He believes that practice and training yield solutions to situations that will inevitably present themselves someday. He calls this his "metaphorical drawer of ideas," and it is a veritable treasure trove.

Abstract reflection pushes Dhont forward in his thinking and calls for an approach very different from that of practical work. He uses drawing to express and form his ideas. His drawing, like his thought, is abstract. The lines and colors represent forms or, more precisely, spaces, along with the two aspects that he believes define a garden's spirit: rhythm and tension. The act of drawing allows him to explore a particular definition of space, to imagine a movement and envision a resolution. His sketches provide him with new ways of creating spaces in a landscape.

This approach is also a device for giving free rein to his ideas, away from material and technical constraints. These eventually come into play in every project, but Dhont considers it important to keep free of them wherever possible, so as not to be trapped in the snare of reality. He is determined not to be restricted by the false constraints of the everyday. It is crucial to him to be able to propose innovative solutions at all times.

Another way in which he expresses his ideas is not graphic but sculptural, and consists of modeling features of a landscape in plaster in order to study the spaces it creates. These models are often produced on a large scale to make it easier to analyze the proportions and make them, quite literally, more palpable. Dhont's landscapes are modeled in every sense of the word. As a result, he often thinks in terms of opposition: positive and negative, empty and full—concepts that frequently recur in his *gypsotheca*, a collection of plaster casts, and which are clearly visible, or legible, in his landscapes on both a large (536 Aalst) and small scale (691 Antwerp).

His sculptural studies include a large collection of topiary shapes designed to work on the reflection of light, the relations between different scales, and the effect of structures and architecture on the perception of space. A piece of topiary, whether in plaster or full size in a garden, makes it possible to understand a number of external qualities, such as the reflection of light on different surfaces, the rhythm of successive planes, or the delineation of a green "room." Other topiary projects are more concerned with organic and tactile qualities (816 La Hulpe) than geometric forms with a strong visual impact (559 Brussels). In both cases, Dhont transforms the way we perceive space by adapting the beauty and craftsmanship of topiary art to contemporary design.

3

1/
793 Sint-Genesius-Rode (BE)
At the edge of a pétanque ground, the existing yew topiary has been integrated into an arrangement of topiary sculpted in cylindrical sections.

2/
691 Antwerp (BE)
Topiary punctuates the scenery in this city garden: sophisticated architectural forms gradually give way to more organic shapes.

3/
557 Saint-Rémy-de-Provence (FR)
These carefully proportioned and positioned box sculptures form a room of greenery that opens onto a magnificent view over the valley.

Dhont often works on fictional and theoretical situations that he has never encountered in real life, and considers this practice crucial to his method. At this early stage, he is dealing with an oneiric garden that exists only in his imagination and in his graphic and sculptural works. Here, everything is possible. Dhont's experiments can move in all directions and on all levels; the sole limits are those of his imagination. It is only later that the studies that make up his metaphorical archives come up against the constraints imposed by the reality of a place and those who live there.

Before this ordeal by fire, inspiration is what counts. It is invariably based on what is already inside the artist—on his or her memory. Without memory, there can be no creation, and so building a rich memory is a way of fostering inspiration. Dhont's research is an act of memory-building. Memory is at the root of his work: each new project sprouts like a young shoot from years of previously accomplished work, much of which is unknown to the public.

Similarly, the finished product, the project, cannot be separated from the process of research and creation. The result is not the only important aspect; the completed project as it is presented to the client also encapsulates the process. This is yet another example of a landscape's temporal aspect, present even before it exists: memory (the result of previous experiments) corresponds to the past, while the creative process takes place in the present and leads to a vision situated in the future.

After drawing and plaster modeling, Dhont proceeds to digitize his projects, working with standard software tools—a procedure also adopted by his team. The next stage is to strike a balance between his vision and the real world's practical limitations.

Even the dream of a landscape ends up taking technical and biological form. As a landscape architect, Dhont is inevitably also an engineer and a botanist. That said, his ultimate goal in creating a landscape is to make people dream, to surprise them, and bring them joy as they move around his microcosms: worlds that always show themselves from their best angles as if it were quite natural.

301 Ohain (BE)

Model for planting *Ginkgo biloba* trees in a picnic area.

620 Brussels (BE)

Architectural domes for an underground swimming pool.

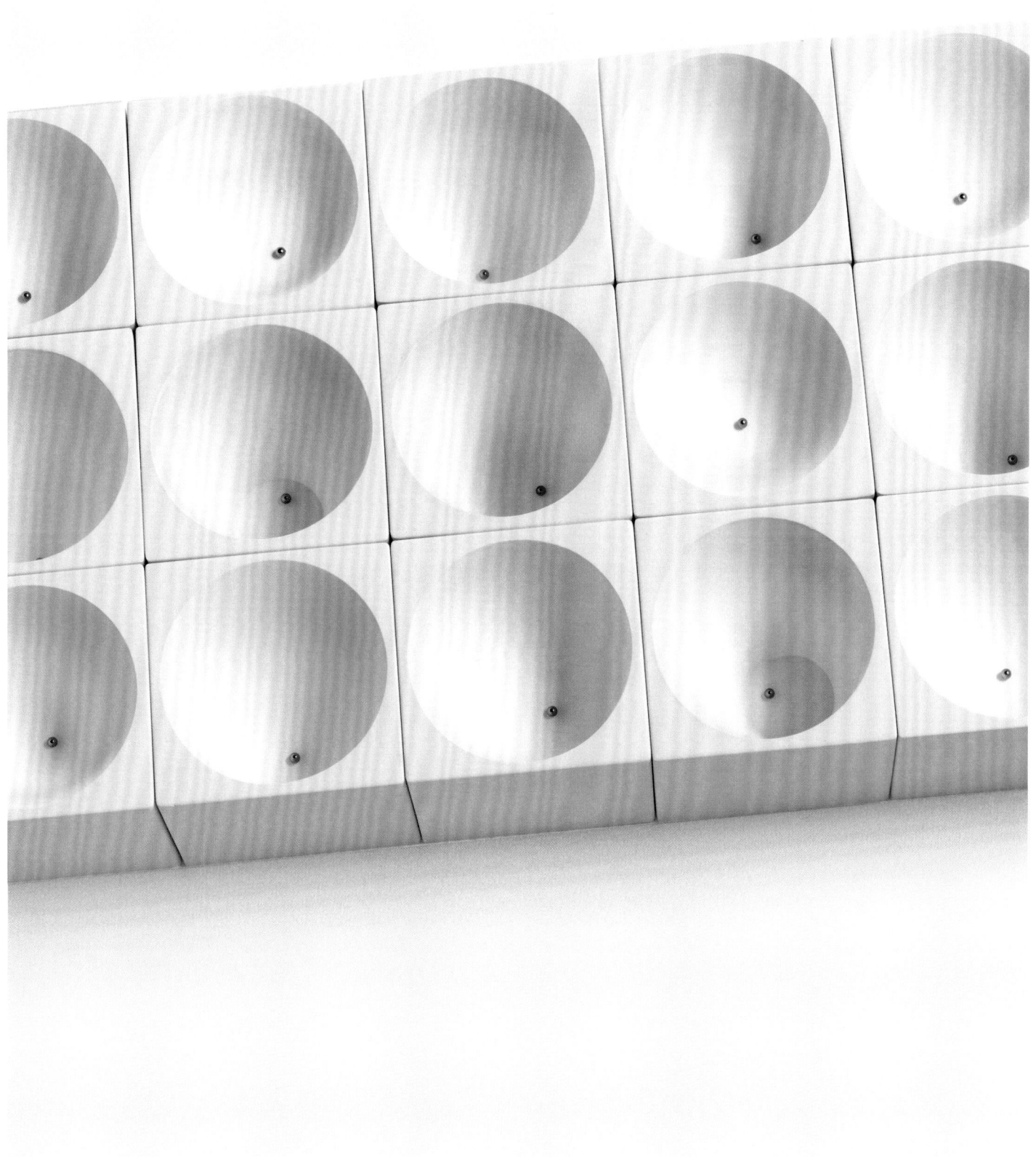

692 Ghent (BE)

Taxus baccata topiaries for the entrance of a temporary exhibition.

Microreliefs for a natural development between geometric paths.

468 Leuven (BE)

Positive and negative landscape for the KU Leuven university campus.

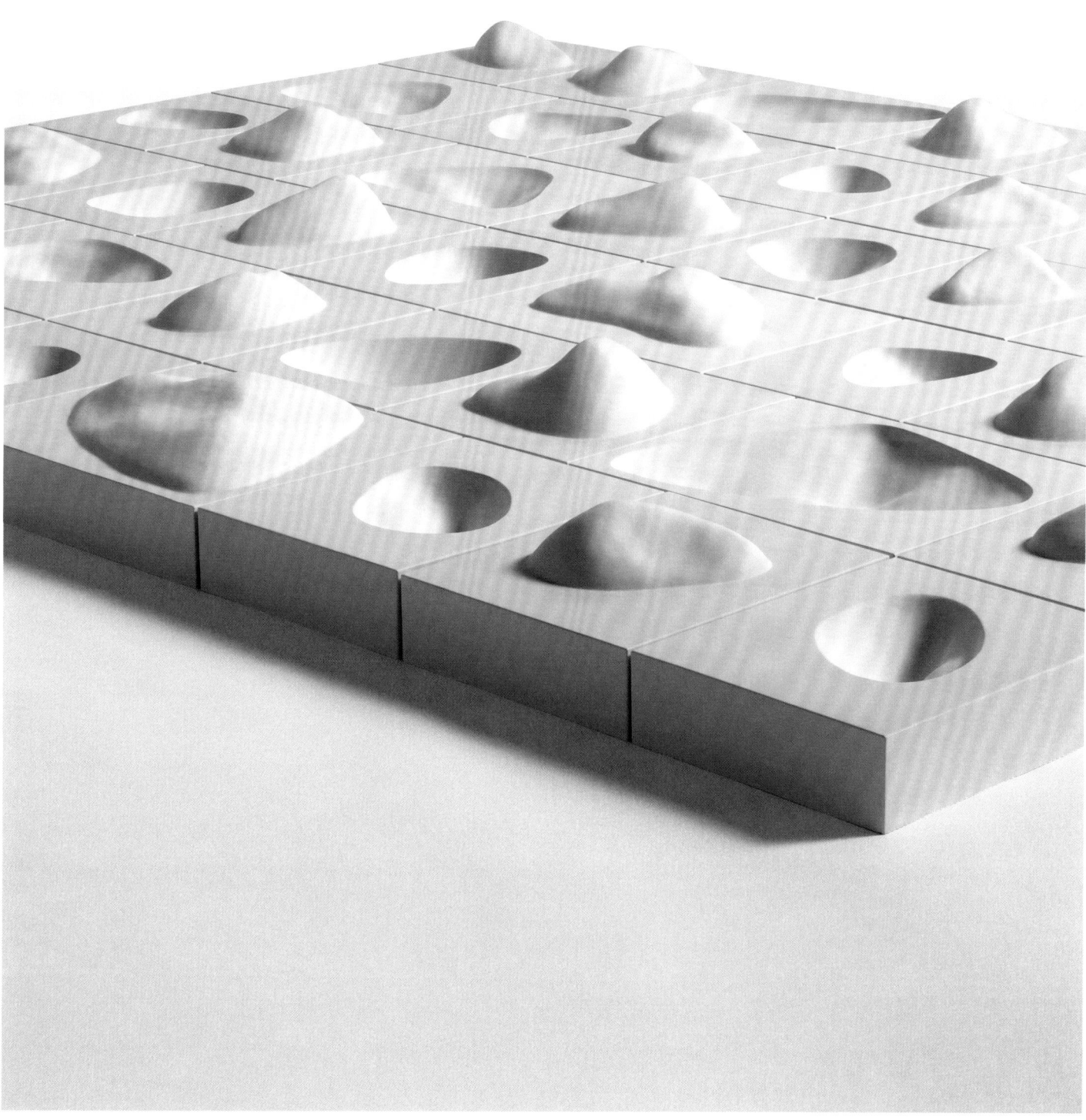

559 Brussels (BE)

Selection of sculptural forms taken from a series of *Carpinus betulus* sculptures.

468 Leuven (BE)

Embankments planted with trees, shrubs, and natural flora with pools of light for an underground parking lot.

Canyon walk in clipped mixed hedges of *Carpinus betulus*, *Fagus sylvatica*, and *Acer campestre*.

584 Oostkamp (BE)

Geometric topiary forms with axis and height differences.

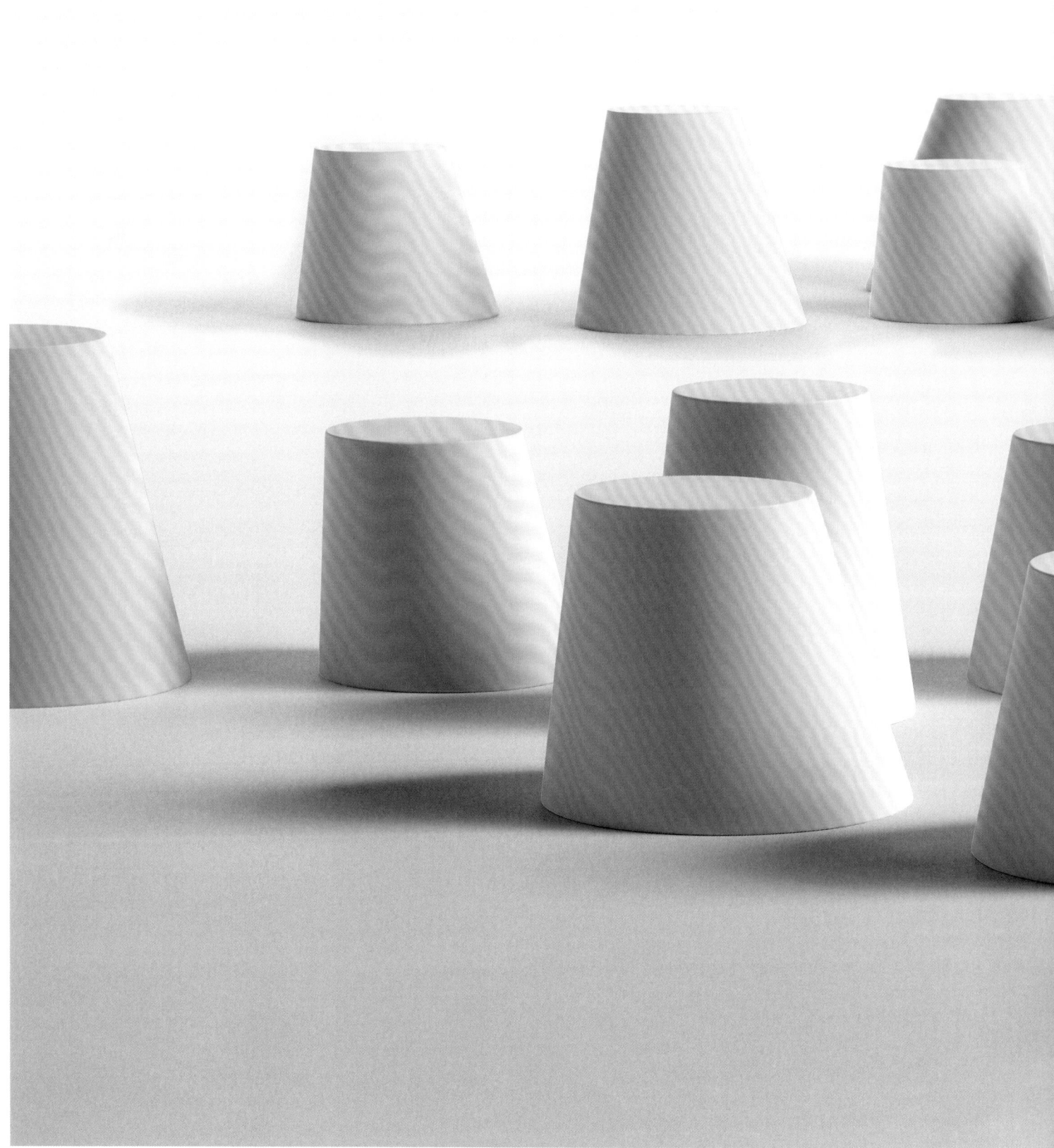

816 La Hulpe (BE)

Taxus baccata clipped as a rough curtain.

691 Antwerp (BE)

Taxus baccata topiaries in a nonarchitectural shape.

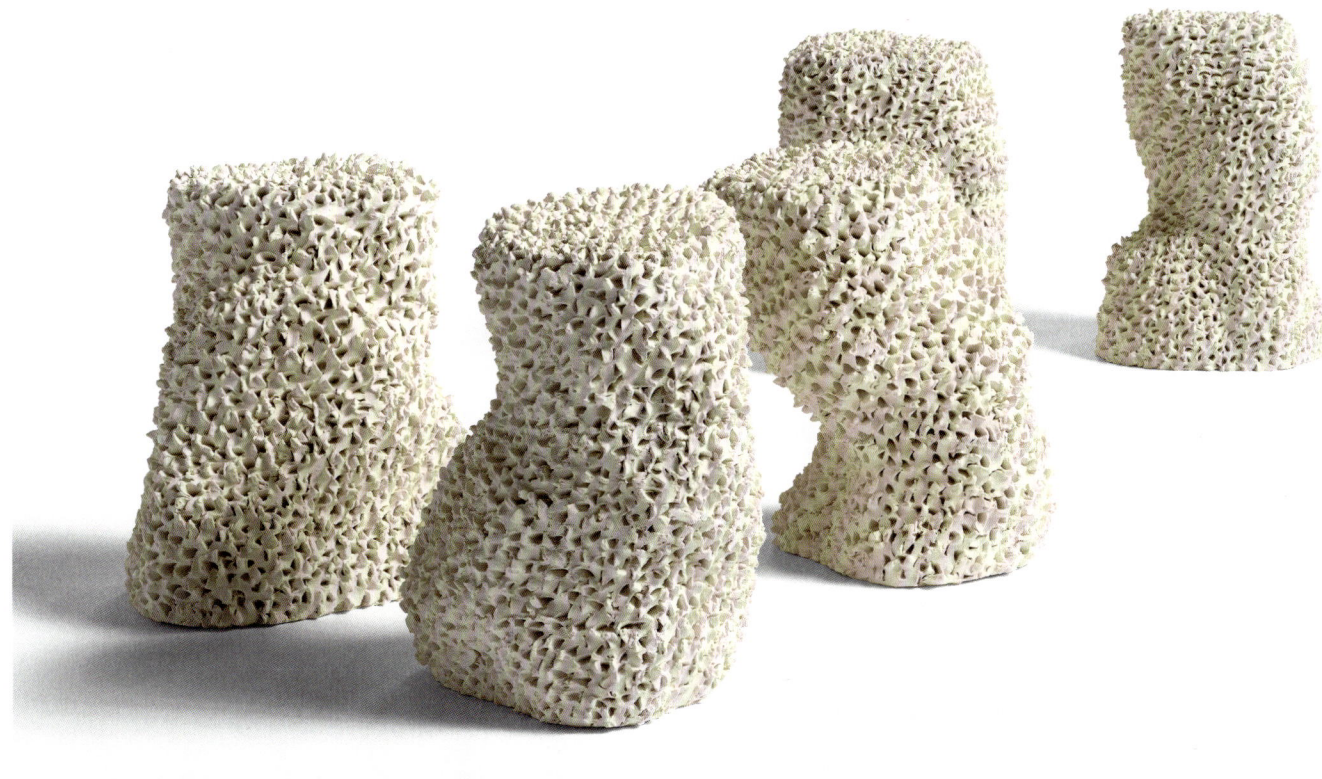

Temporary green topiaries for a public square.

Ode to the Earth,
Ode to the Landscape

Erik Dhont

A lifespan is too short to understand nature in the abstract; we learn about it through experience. On another level, the knowledge and passion of a patron inspire reflection and contribute to our understanding of nature and humankind.

A landscape project is above all the product of observation, of memories seen through a filter that isolates and distills the real-world elements surrounding us. The resulting frame of reference gives us freedom to organize our thoughts, because it allows us to think outside the immediate context and brings us to the threshold of the unreal and inexpressible. This process is the life force that enables each project to break free from the prevailing conformism.

Chia & Florence (IT)

The symbiosis of nature and architecture can assume surprising forms. Take, for example, the Tower of Chia in the community of Soriano in Italy—the remains of a medieval castle, around which a substantial copse has grown over the years. From the surrounding area, the sight of the tower as it emerges from the wild thicket has an uncanny appeal, and yet no one planned this tension between nature and architecture.

Chia (IT) – A glimpse of the tower through a cluster of trees.   Florence (IT) – A solitary walkway beneath the laurels of Boboli.

Elsewhere, the relationship between buildings and nature has been carefully thought out. Another essential point of reference, also in Italy, is the long arbor of *Laurus nobilis* known as the *cerchiata* (circular) in the Boboli Gardens in Florence. A solitary walk in this elegantly simple leafy corridor is an unforgettable experience. Beyond personal discovery, though, such places are also important for the way in which they establish a free space for appropriation—of solitude as a retreat or refuge, or a more political form of appropriation triggering social interaction.

Vaux-le-Vicomte (FR) & Hemelrijk (BE)

As designers, we search for a through-line to navigate free space by working on both the horizontal and vertical plane, and using living and inert material. The orchestration of space and the sequence of atmospheres are focused around pivotal points central to the design. At Vaux-le-Vicomte, the rhythm and density of the planting scheme make for a sophisticated intimacy before the garden opens onto the wide vista that is the promenade's principal feature. The layout of the individual elements make this zone appear longer than it is. To understand the importance of the spaces, it helps to appreciate the significance of emptiness, because our perception of it determines the way we see them. The right balance between open space and built or planted space creates a sense of equilibrium.

Vaux-le-Vicomte (FR) – Opening the space accentuates the size of the buildings.

Hemelrijk (BE) – Russell Page's botanical combinations give structure to the garden.

Something else to strive for is a sense of wonder. The botanical or spatial surprise evoked in us as we walk through a garden may thus be a result of the landscape architect's game, stirring flashes of unexpected emotion.

A visit to the rose breeder Louis Lens elicits a moving encounter with the early-flowering Albertine rose. This plant flowers only once a year and, considering the weather's unpredictability and other natural constraints, the suspense of waiting for it to bloom intensifies pleasure when it does. A garden can be seen as the frame for a moment of wonder, and its memory makes us look forward to upcoming seasons. We might, at any time, be surprised by a new blossom or a certain light or atmosphere: in a garden, every moment is unique. A work by Russell Page in Essen combines *Cryptomeria japonica* and witch hazel around a circular pool of water. The sight of this garden in the snow when the witch hazel is blooming etches itself on the memory.

The appreciation of such moments is not, however, absolute, but depends on certain paradigms that influence the way we see, just as our notions of garden and landscape are culturally determined. Our relationship with nature has been reinterpreted time and again through various aesthetic and scientific revolutions. Christian Cay Lorenz Hirschfeld, an Enlightenment theorist and the author of *Theory of Garden Art* (1775-77), clearly explains the cultural construction and codes necessary to understanding the art of gardening. In his book he defends the notion that a landscape can be interpreted from a philosophical, political, or moral perspective.

Rosa "Albertine" is a sublime scented pink rose.

Christian Cay Lorenz Hirschfeld's epitaph: a memorial, and an ode to nature.

At Vaux-le-Vicomte, the garden is a demonstration of individuality and personality. Humanity assumes the role of master of nature, manifesting itself in the concrete form it gives its intentions, power, and knowledge.

In Jean-Jacques Rousseau, on the other hand, the garden is a place where nature and humanity are close, and the relationship between them can be explored. In his epistolary novel *La Nouvelle Héloïse* (*Julie; or, The New Heloise*), Julie's secret garden is a key site where a new way of life is revealed, offering a concrete alternative to the complex triad of love, individuality, and society.

More generally, the second half of the eighteenth century marked an important turning point in our relationship with nature, which ceased to be a cause of fear and began to be an object of individual and imaginative exploration. Garden art and the perception of landscape were transformed as a result.

Brussels (BE) & Geneva (CH)

Today we can create a frame for our lives by delineating a chosen field of vision. Here, architecture proceeds by contrast. Like a stage set, it takes second place to content, and nature becomes a living painting. This focus on the garden encourages contemplation, highlighting its various features and accentuating the effects of each season.

Brussels (BE) – The panoramic openings in the Van Buuren Museum frame the views of the garden.

Geneva (CH) – A clearing in a romantic oak grove.

Perhaps there is a tendency at present to attach more importance to observing nature than to taking action on it; though not inactive, we prefer to give free rein to the diversity of the surrounding plant world. We are attracted by the sight of a landscape where colors, textures, and weather are clearly conveyed. As in Rousseau's novel, the richness lies perhaps in a heightened and awed awareness of the botanical world and the merits of nature. Following in the footsteps of the philosopher on the road to paradise is as much a pilgrimage as a learning process. And on the subject of learning, what is the best training for gardeners and landscapers? Should they sketch places out of context and speculate on potential constraints to their projects? Should they study the spatial aspects of famous historic monuments? Or would it be better if they immersed themselves in researching local sensibilities and culture? Finally, isn't it important not only to observe, but also to take action—to prune the tree that is too tall and sow the seeds that will bear fruit?

Gravetye Manor (GB) & Mildam (NL)

One possibility would be to follow in the footsteps of William Robinson, a gardener who cultivated the wild side of his creations. In the wooded gardens he designed in

Britain in the late nineteenth century, there is an emphasis on certain natural features. It would also be possible to go further and highlight natural processes over features. Louis le Roy's eco-cathedrals in the Netherlands are a good example of such an approach.

For a landscape architect, the challenge is to create long-term added value. This value is based on an underlying social vision, but because concepts can repeat themselves over time, it must be given concrete form in a spatial structure. People must be encouraged to look at the world from different angles and—perhaps inspired by an awareness of botany—to inhabit outside space in a different way.

Spatial scenery, contouring, and botanical selection form the basis of research that leads to the creation of places in which life is given free rein. The most successful among them will, paradoxically, not feel planned, but as if they had always existed.

East Grinstead (GB) – The wild manor garden created by William Robinson.

Mildam (NL) – Louis le Roy's eco-cathedrals, deep in the countryside.

It would be possible to mention many further examples and attempt to explain the striking beauty of the moments we spend in contact with nature. As I am not a theorist, however, I shall restrict myself to celebrating the observation and love of our environment.

Every garden created is a place for representing values and experimenting with human relationships that are anchored in space and time, transmitted, and transformed. This initiatory site is the expression of a universal quest: the search for a world view that makes sense. Humans need the earth. Humans should plant, and keep planting forever. By building an intimate relationship with nature and plants, all humans can develop their aesthetic and ecological sensitivity and become aware of the countless potential relationships the world offers; subtle relationships that hold the seeds to future landscapes.

Site Plans

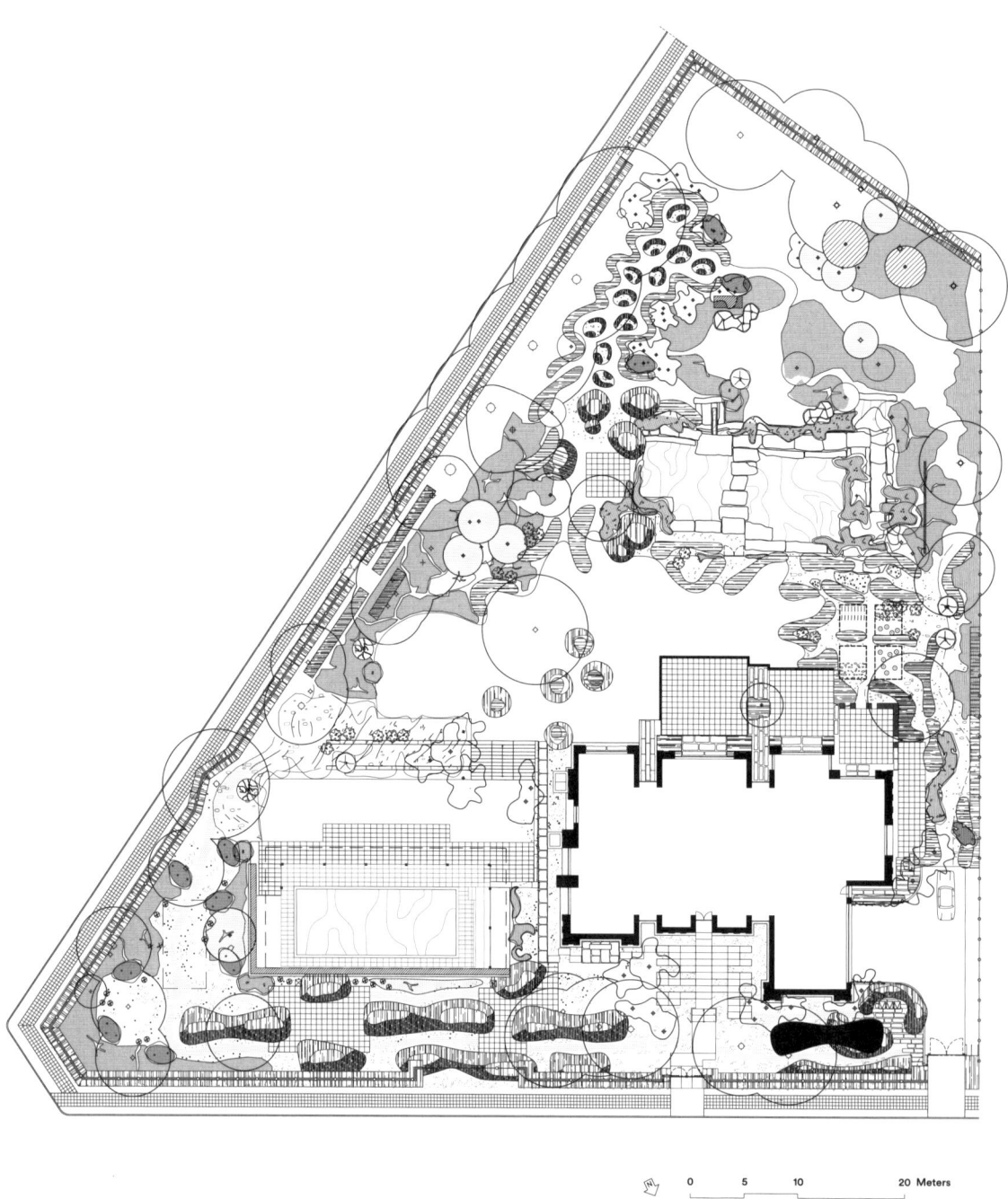

0   5   10   20 Meters

0   5   10   20 Meters

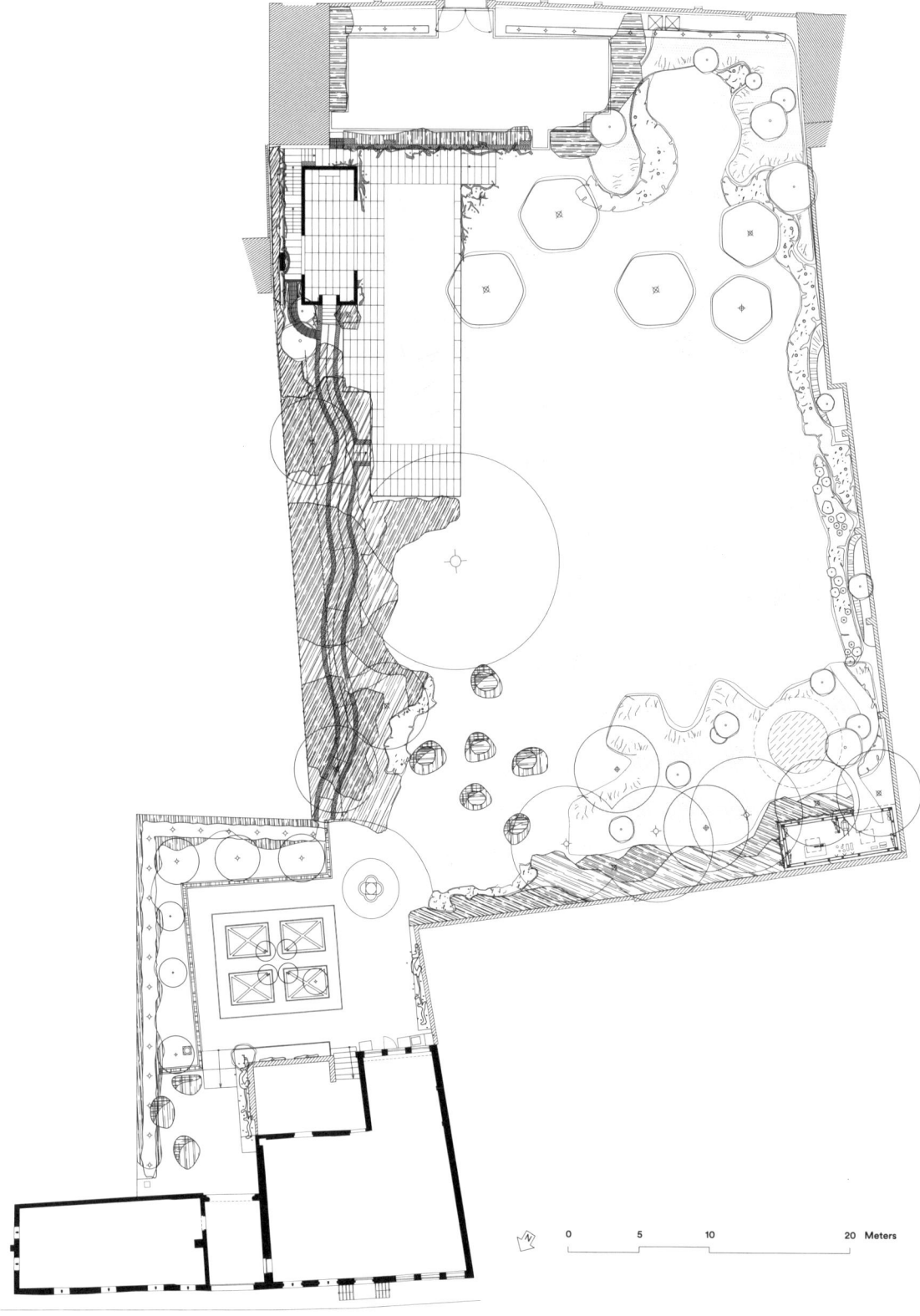

0   5   10   20 Meters

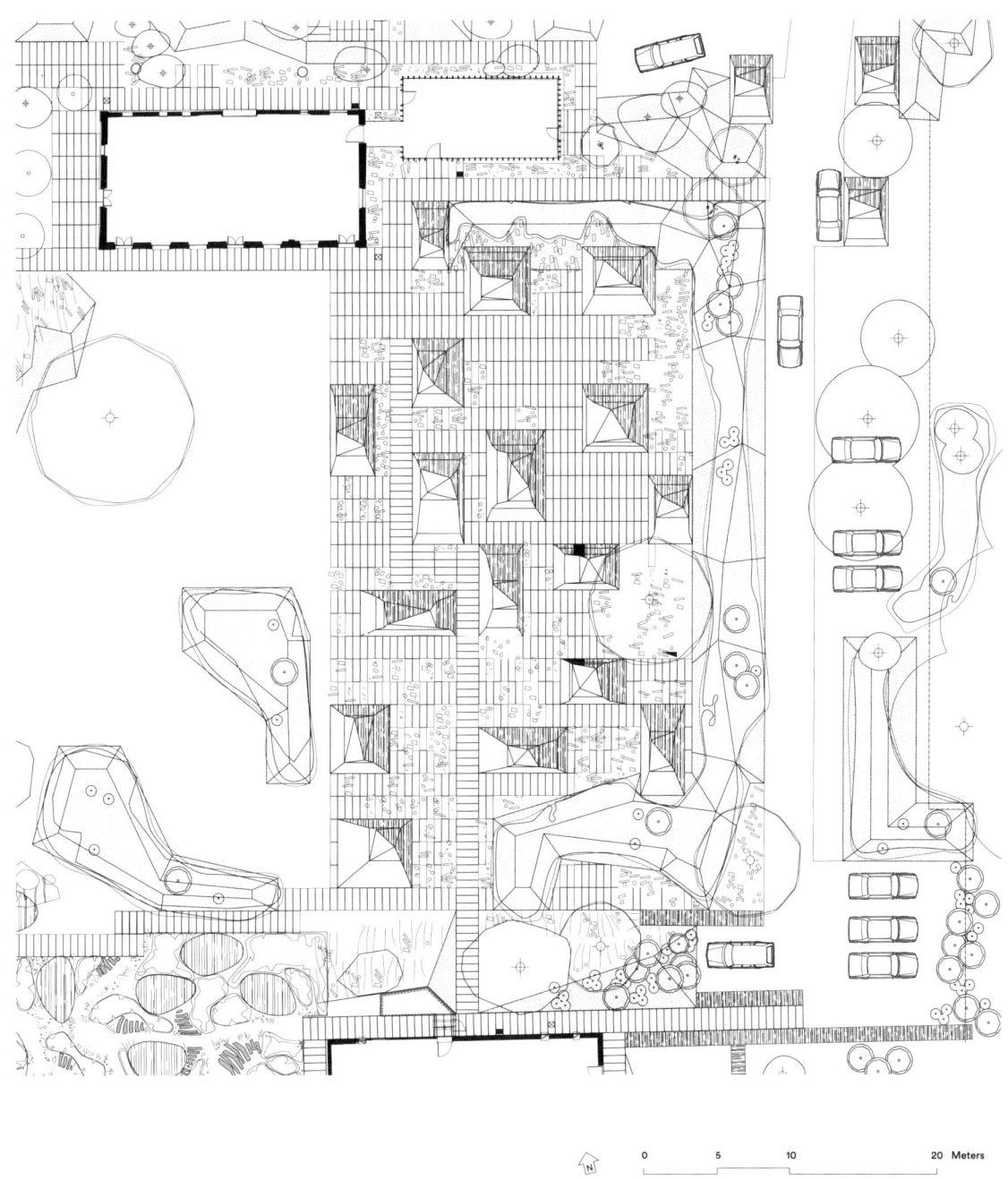

0   5   10   20 Meters

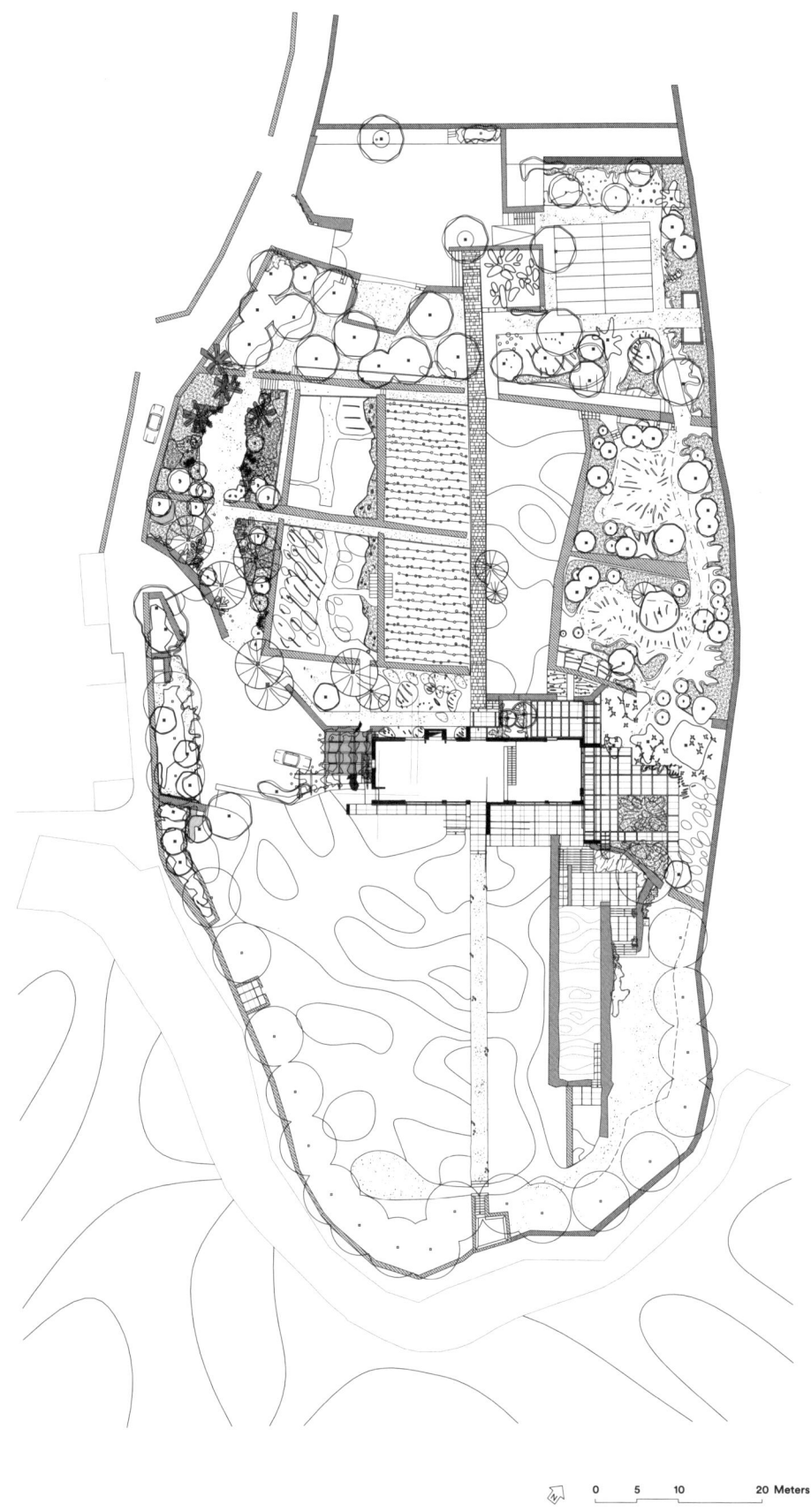

0   5   10   20 Meters

0   5   10   20 Meters

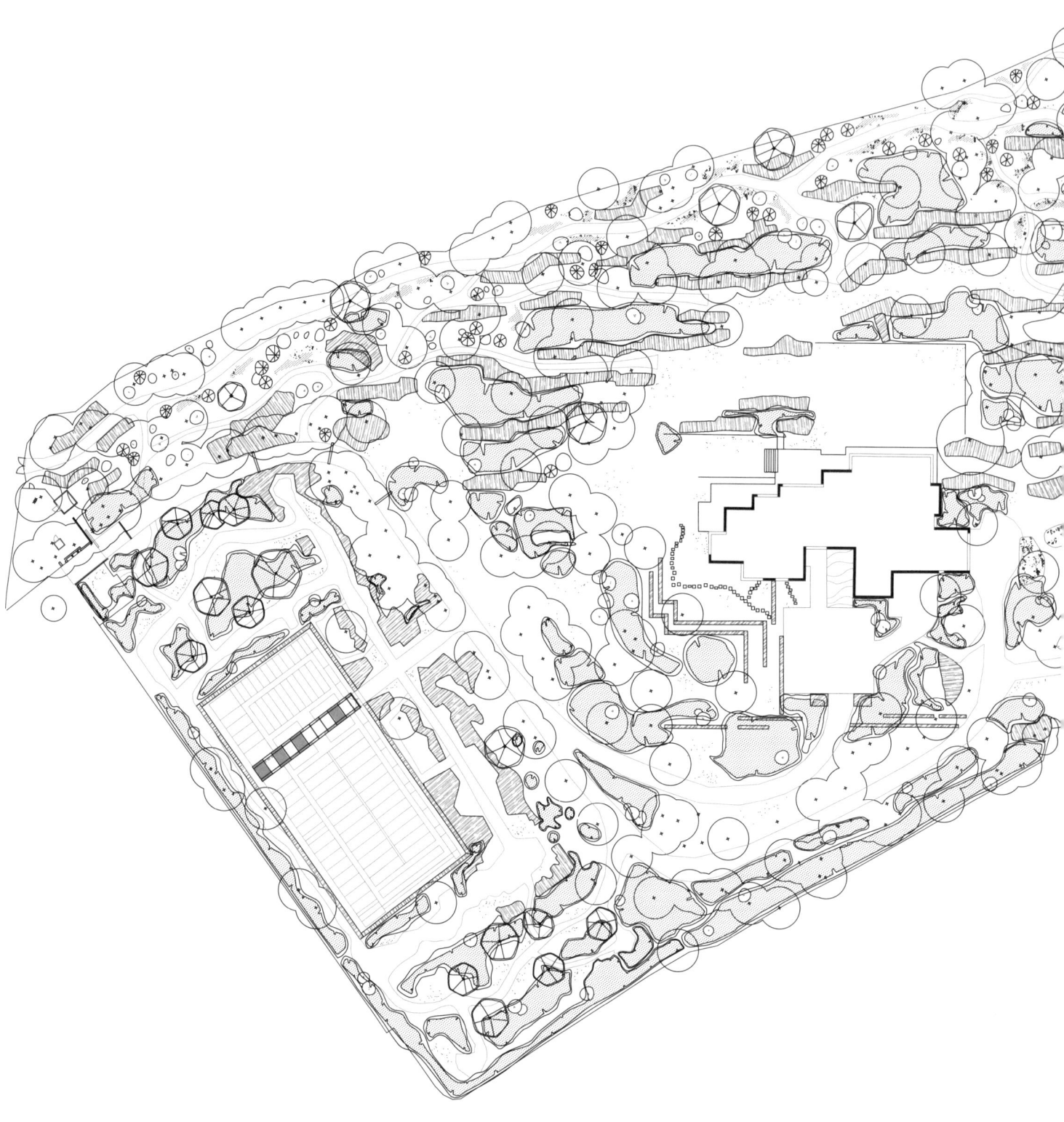

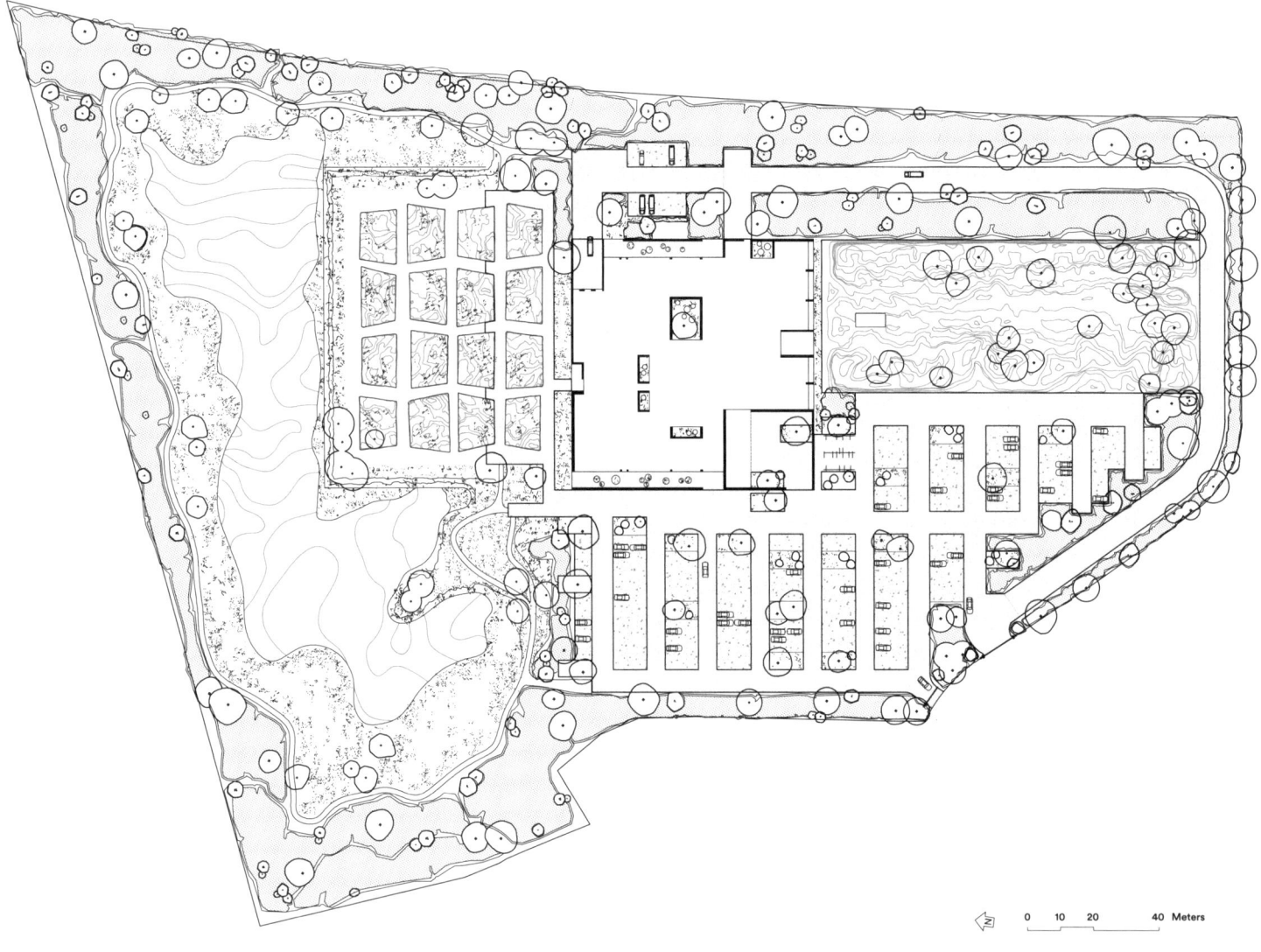

0   10   20   40 Meters

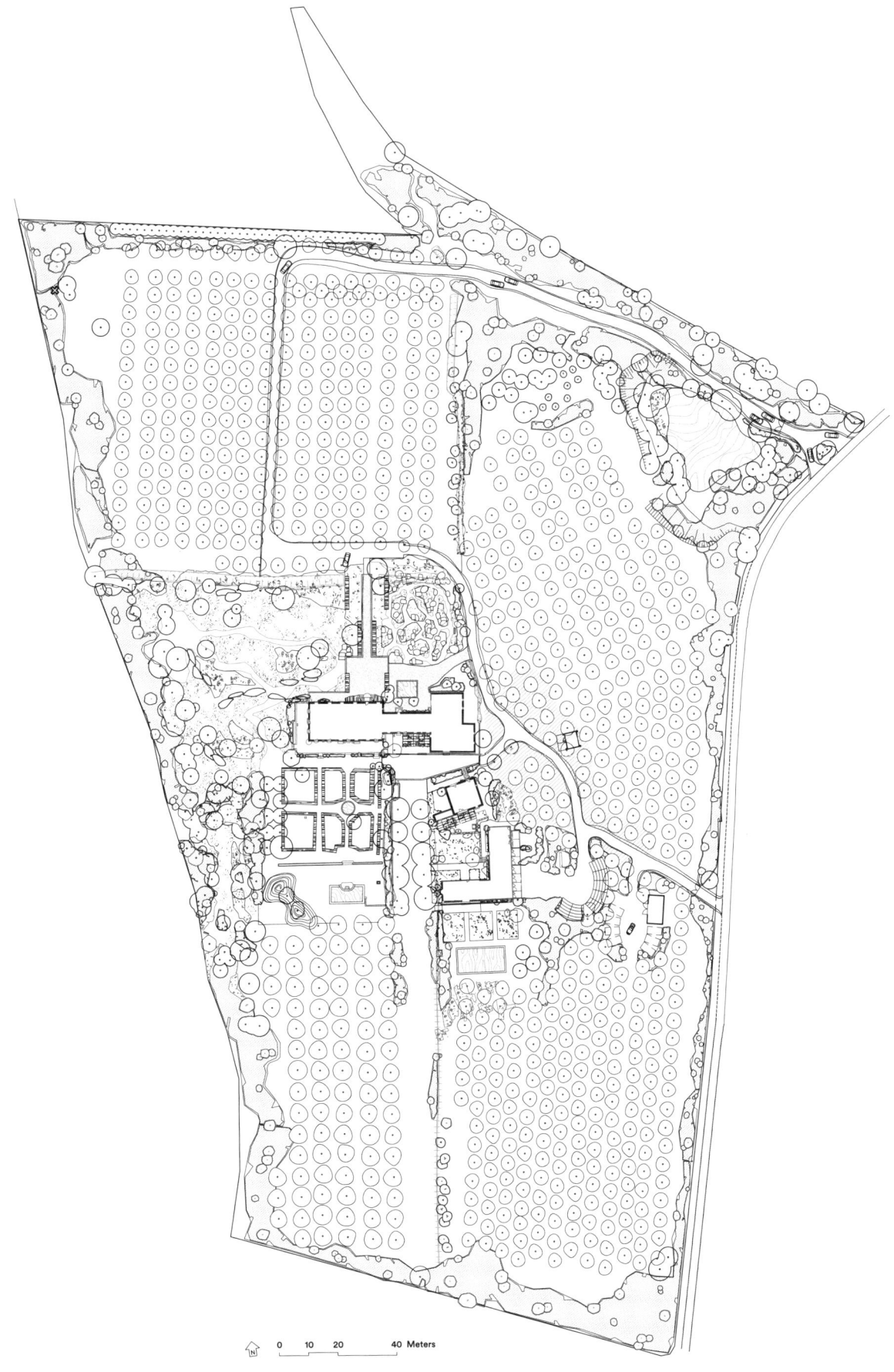

0   10   20   40 Meters

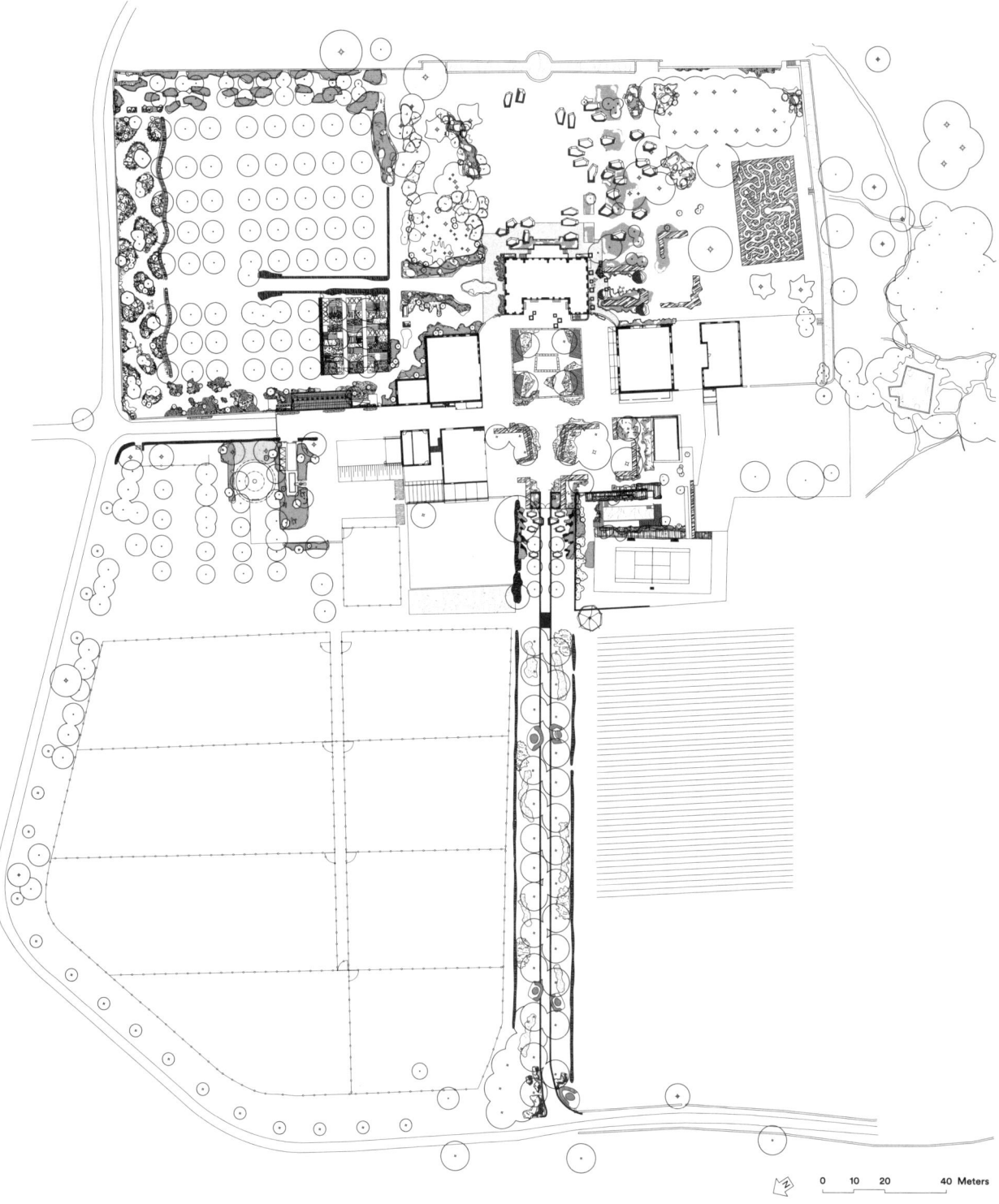

0 10 20   40 Meters

Present collaborators

Tom Baelus
Lucie Bécu
Jules Bongrand
Florian Boniface
Niels Everaerd
Tim Vaculik

Former collaborators

Branislav Andel
Claudia Avila Gomes
Daphnis Bockstael
Frederik Cassiman
Mehdi Delporte
Lorenzo de Simone
Bruno Eeman
Elvire Evrard
Tom Feyen
Catherine Fierens
Wim Gerolt
Joseph Ingenito
Abraham Morcillo
Nicolas Pauwels
Floris Steyaert
Kenneth Van der Taelen
Pieter Van Hauwermeiren
Maarten Vansteenhuyse
Winifred Van Wonterghem

Consultants

Francis Adams
Marc Appelmans
Laetitia Fraikin

Acknowledgements

Erik Dhont would like to express his sincere thanks
to Ms. Nathalie and Mr. Simon Marshall Lockyer,
and to Ms. Verena and Mr. Rémy Best,
who made this book possible.

Erik Dhont Landscape Architects
30 Marché aux Porcs, 1000 Brussels
+32 2 219 19 02
info@erikdhont.com
www.erikdhont.com

Editors: Erik Dhont, Suzanne Kříženecký
Authors: Erik Dhont, Michael Jakob,
    Suzanne Kříženecký
Project management: Florian Boniface (Erik Dhont
    Landscape Architects), Dorothee Hahn (on behalf of
    Hatje Cantz)
Copyediting: Kimberly Bradley
Translation from the French: Imogen Taylor
Graphic design: Jurgen Persijn (N.N.)
Reproductions: DLG Graphic, Paris
Production: Thomas Lemaître (Hatje Cantz)
Paper: Gardapat Bianka, Munken Lynx rough
Printing and binding: Printer Trento s.r.l.

Published by
Hatje Cantz Verlag GmbH
Mommsenstraße 27
10629 Berlin
Germany
www.hatjecantz.com

A Ganske Publishing Group Company

ISBN: 978-3-7757-4815-5